Emerging Critical Technologies and Security in the Asia-Pacific

Emerging Critical Technologies and Security in the Asia-Pacific

Edited by

Richard A. Bitzinger
Senior Fellow, Nanyang Technological University, Singapore

First published 2016 by
PALGRAVE MACMILLAN

Palgrave Macmillan in the UK is an imprint of Macmillan Publishers Limited,
registered in England, company number 785998, of Houndmills, Basingstoke,
Hampshire RG21 6XS.

Palgrave Macmillan in the US is a division of St Martin's Press LLC,
175 Fifth Avenue, New York, NY 10010.

Palgrave Macmillan is the global academic imprint of the above companies
and has companies and representatives throughout the world.

Palgrave® and Macmillan® are registered trademarks in the United States,
the United Kingdom, Europe and other countries.

ISBN 978–1–137–46127–8

This book is printed on paper suitable for recycling and made from fully
managed and sustained forest sources. Logging, pulping and manufacturing
processes are expected to conform to the environmental regulations of the
country of origin.

A catalogue record for this book is available from the British Library.

Library of Congress Cataloging-in-Publication Data
Bitzinger, Richard.
    Emerging critical technologies and security in the Asia-Pacific /
Richard A. Bitzinger, Senior Fellow, Nanyang Technological University,
Singapore.
      pages   cm
    Includes index.
    ISBN 978–1–137–46127–8 (hardback)
    1. Military art and science—Technological innovations—Asia.
    2. Military art and science—Technological innovations—Pacific Area.
    3. Security, International—Asia.   4. Security, International—Pacific
    Area.   I. Title.
    U43.A8A78 2016
    355′.03305—dc23   2015032820

Contents

Figures and Tables

Figures

Tables

Acknowledgments

This volume could not have been possible without the considerable support and inputs of many talented individuals and their respective organizations. In the first place, the research that went into this book was only possible due to a generous grant from the John D. and Catherine T. MacArthur Foundation, through its "Asian Security Initiative" program, made to the S. Rajaratnam School of International Studies (RSIS) at the Nanyang Technology University. Additionally, I am grateful to the management at RSIS—particularly Dean Barry Desker and Dr. Tan See Seng, Director and Head of Research at RSIS's Institute of Defense and Strategic Studies—who gave this project their full and enthusiastic support. Thanks are also due to the Events Unit at RSIS who organized and ran an excellent workshop that in turn produced most of the chapters found in the book. Finally, I am grateful to all of the writers who contributed to this volume, and, last, but most certainly not least, to the continuing support of my family, particularly my daughters, Jennifer and Amy.

R.A.B., Singapore

Contributors

Kogila Balakrishnan is a principal fellow at WMG, University of Warwick. She holds a PhD from the Defense College of Management and Technology, Cranfield University. She is also a visiting fellow at the Cranfield University and adjunct professor at the National Defense University, Malaysia. Her research interest is on industry and technology policy, defense procurement, offsets and impact of technology transfer. She has held research positions at the Centre for Defense Acquisition, Cranfield University at the UK Defence Academy, and was the Under Secretary of the Defense Industry Division of the Malaysian Ministry of Defense. She has authored several articles, and book chapters, including "Malaysia's Defense Industry and the Role of Offsets," *Journal of Defense Peace and Economics* (August 2009), and "International Offsets Experience and Policy Prescription," in *Defense Acquisition, International Best Practices* (2013).

Richard A. Bitzinger is a senior fellow and coordinator of the Military Transformations Program at the S. Rajaratnam School of International Studies (RSIS), Nanyang Technological University (Singapore), where his work focuses on security and defense issues relating to the Asia-Pacific region, including military modernization and force transformation, regional defense industries and local armaments production, and weapons proliferation. He has written several monographs and book chapters, and his articles have appeared in such journals as *International Security, Orbis, China Quarterly*, and *Survival*. He is the author of *Towards a Brave New Arms Industry?* (2003), "Come the Revolution: Transforming the Asia-Pacific's Militaries," *Naval War College Review* (Fall 2005), *Transforming the U.S. Military: Implications for the Asia-Pacific* (December 2006), and "Military Modernization in the Asia-Pacific: Assessing New Capabilities," *Asia's Rising Power* (2010). He is also the editor of *The Modern Defense Industry: Political, Economic and Technological Issues* (2009). He was previously an associate professor with the Asia-Pacific Center for Security Studies (APCSS), Honolulu, Hawaii, and has also worked for the RAND Corporation, the Center for Strategic and Budgetary Affairs, and the US government. In 1999–2000, he was a senior fellow with the Atlantic Council of the United States.

Eugene Gholz is an associate professor at the Lyndon B. Johnson (LBJ) School of Public Affairs, University of Texas Austin. He works primarily at the intersection of national security and economic policy, on subjects including innovation, defense management, and US foreign policy. From 2010 to 2012, he served in the Pentagon as Senior Advisor to the Deputy Assistant Secretary of Defense for Manufacturing and Industrial Base Policy, where he led initiatives to better understand the complex defense supply chain and on reimbursement of industry's Independent Research and Development (IR&D) expenditures. Before working in the Pentagon, he directed the LBJ School's master's program in global policy studies from 2007 to 2010. He is the coauthor of *Buying Military Transformation: Technological Innovation and the Defense Industry*, and *U.S. Defense Politics: The Origins of Security Policy*. His recent scholarship focuses on energy security. He is also a research affiliate of Massachusetts Institute of Technology's (MIT) Security Studies Program, a member of the Council on Foreign Relations, and associate editor of the journal *Security Studies*. His PhD is from MIT.

Caitríona Heinl is a research fellow responsible for cybersecurity research at the Centre of Excellence for National Security, RSIS, Singapore. She is a UK-trained solicitor (non-practicing) and admitted as an Attorney-at-Law in New York. She was previously responsible for Justice and Home Affairs policy and the Justice Steering Committee at the Institute of International and European Affairs (IIEA), Ireland, where she analyzed European and international criminal justice issues. She was the legal researcher and IIEA-based project manager on a study for the European Commission's Directorate-General Justice, Liberty and Security on non-legislative measures to prevent the distribution of violent radical content on the Internet. Her speaking engagements include the ASEAN Regional Forum (ARF) Workshop on Cyber Confidence Building Measures; the NATO Cooperative Cyber Defense Center of Excellence (CCDCOE) annual cyber conflict conference; and the Network of ASEAN Defense and Security Institutions Workshop on Cybersecurity. She holds an MPhil in international relations from the University of Cambridge.

Andrew James is a senior lecturer at Manchester Business School and a senior research fellow at the Manchester Institute of Innovation Research. He is the author or co-author of more than 50 academic journal publications, book chapters, and public domain reports and his research focuses on the application of core frameworks from the academic literature on technology strategy and innovation management to the particular challenges and particularities of the security and defense sectors.

Kalyan M. Kemburi is an associate research fellow with the Military Transformations Program at RSIS, Singapore. He has previously worked with the James Martin Center for Nonproliferation Studies, California, and the UN Office for Disarmament Affairs, New York. He has published with the *Bulletin of the Atomic Scientists* and the *Nuclear Threat Initiative* as well as contributed chapters to volumes published by Center for Strategic and International Studies (CSIS), Carnegie Endowment, Oxford University Press, and World Scientific. Along with Li Mingjiang from RSIS, he is working on two edited volumes, *China's Power and Asian Security* and *New Dynamics in US-China Relations*. He holds a master's in international policy studies and Certificate in Nonproliferation Studies from the Monterey Institute of International Studies, and an MSc in international political economy (recipient of the 2010 Lion Group Gold Medal) from the Nanyang Technological University, Singapore.

Bernard F.W. Loo is Associate Professor and Coordinator of the Master of Science in Strategic Studies degree program at RSIS, Nanyang Technological University. He is the author of *Medium Powers and Accidental Wars: A Study in Conventional Strategic Stability* (2005) and editor of *Military Transformation and Operations* (2009). His other publications have appeared in the *Journal of Strategic Studies, Contemporary Southeast Asia, NIDS Security Reports*, and *Taiwan Defense Affairs*. He is a regular commentator on defense matters, both in Asian newspapers and in defense-related institutions and conferences in Europe and Asia.

Martin Lundmark is deputy research director at the Swedish Defense Research Agency (FOI). He holds a PhD in business administration from the Stockholm School of Economics. The dissertation title was "Transatlantic Defense Industry Integration—Discourse and Action in the Organizational Field of the Defense Market." He has for 16 years studied the international defense industry, acquisition, innovation, and research. His present research primarily concerns national defense innovation systems. He has been a guest researcher at MIT and the Foundation for Strategic Research (FRS) in Paris. He presently collaborates with a number of internationally renowned research centers and universities, including RSIS, the Universidade Federal Fluminense (Brazil), FRS, and the University of Vancouver.

Paul T. Mitchell has worked at the Canadian Forces College since 1998, first as the Deputy Director of Academics, and later as its first Director of Academics (DAcad). As DAcad, he oversaw the development of the Master of Defense Studies degree following accreditation of the Command and Staff Course by the Ontario Council of Graduate Studies in 2001. Between 2005 and 2007, he was an associate professor at Singapore's RSIS, Nanyang Technological University and also taught at the Singapore Armed Forces Training Institute's Command and Staff College. He holds a PhD from Queen's University, Kingston. Following the completion of his doctoral studies, he worked as a post-doctoral fellow at Dalhousie University in Halifax in the Centre for Foreign Policy Studies, where he assisted with the production of the Canadian Navy's "Adjusting Course" strategy. He also worked as Directing Staff at the Pearson Peacekeeping Centre on their Maritime Peacekeeping course in 1996 and 1997. In 2003, he was awarded the Literary Award by the United States Naval Institute and the Surface Naval Association for his paper "Network-Centric Warfare and Small Navies"; he was the first non-American and the first civilian to be so recognized.

Michael Raska is a research fellow in the Military Transformations Program at RSIS, Nanyang Technological University, Singapore. His research interests encompass East Asian security and defense, including theoretical and policy-oriented aspects of military innovation, force modernization trajectories, strategic studies, and future of warfare. He has published on these issues in the *Korean Journal of Defense Analysis*, *Porinter—The Journal of the Singapore Armed Forces*, and *The Asian Journal of Public Affairs*, and he has contributed chapters to several edited volumes. He has taught at the Goh Keng Swee Command and Staff College and the Lee Kuan Yew School of Public Policy. His research experiences include visiting fellowships at the Hebrew University of Jerusalem, Yonsei University, Pacific Forum CSIS, and the Samsung Economic Research Institute. He holds a PhD in public policy (2012) from the Lee Kuan Yew School of Public Policy, where he was a recipient of the NUS President's Graduate Fellowship.

Andrew L. Ross is Professor of International Affairs at the George H.W. Bush School of Government and Public Service, Texas A&M University. From 2005 to 2014, he was Professor of Political Science, University of New Mexico, where he also served as Director of the Center for Science, Technology, and Policy and Director of Special Science, Engineering, and Policy Research Initiatives in the Office of the Vice President for

Research. He previously served on the faculty of the US Naval War College for 16 years. His work on US grand strategy, national security and defense planning, regional security, and security and economics has appeared in numerous journals and books. His work focuses on the US grand strategy debate, military innovation, and nuclear policy, strategy, and force structure. He holds an MA and a PhD from Cornell University and a BA from American University and has held fellowships at Cornell, Princeton, Harvard, and the University of Illinois.

Virginia Watson is an associate professor at the Asia-Pacific Center for Security Studies in Honolulu, Hawaii. Her areas of interest include science, technology, and security in the Asia-Pacific region, water security, and Southeast Asia geopolitics. She has held appointments at the University of Denver and Colorado School of Mines, and served as an exchange faculty for the University of Colorado in Beijing, China, and is a consultant for various institutions. She is working on *Governance in Asia: Issues in Emerging Technologies* and is editor of *Science, Technology and Security: Issues of Concern Now and Ahead*. She holds a master's in Asian studies from Cornell University and a PhD in international studies in public policy and international technology assessment and management from the University of Denver's Josef Korbel School of International Studies.

Introduction

Richard A. Bitzinger

Technology is widely regarded to be a crucial element of military effectiveness and advantage. As Keith Krause put it, "the possession of modern weapons is a key element in determining the international hierarchy of power."[1] In theory (and often in practice), the possession of cutting-edge militarily relevant technologies equals more effective weapons systems, which results in greater military power, which in turn translates into greater geopolitical power. At the same time, the transnational diffusion of military-related technologies is an important factor affecting the distribution of power in international politics.[2] Consequently, the global dissemination of advanced, militarily relevant technologies should be as great a security concern as the spread of weapons systems themselves.

Regarding the Asia-Pacific, the proliferation of advanced conventional weaponry to this region over the past two decades or so has been momentous, and perhaps even alarming. Since the 1990s, Asian-Pacific militaries have greatly expanded their warfighting capacities beyond the mere modernization of their armed forces, that is, simply replacing older fighter aircraft with more sophisticated versions, or buying new tanks and artillery pieces. In fact, many militaries in the region have over the past 20 years added capabilities that they did not possess earlier, such as new capacities for force projection and stand-off attack, low observability (stealth), and greatly improved command, control, communications, computing, intelligence, surveillance, and reconnaissance (C4ISR) networks. At the very least,

The acquisition of these new military capabilities has many implications for militaries in the Asia-Pacific. [They] promise to upgrade and modernize war fighting in the region significantly. Certainly,

1

Asia-Pacific militaries are acquiring greater lethality and accuracy at greater ranges, improved battlefield knowledge and command and control, and increased operational maneuver and speed. Stand-off precision-guided weapons, such as cruise and ballistic missiles and terminal-homing (such as GPS or electro-optical) guided munitions, have greatly increased combat firepower and effectiveness. The addition of modern submarines and surface combatants, amphibious assault ships, air-refueled combat aircraft, and transport aircraft have extended these militaries' theoretical range of action. Advanced reconnaissance and surveillance platforms have considerably expanded their capacities to look out over the horizon above, below, and on the sea surface. Additionally, through increased stealth and active defenses (such as missile defense and longer-range air-to-air missiles), local militaries are adding substantially to their survivability and operational effectiveness.[3]

Consequently, "conflict in the region, should it occur, would likely be more 'high-tech' than in the past—faster, longer in reach, and yet more precise and perhaps more devastating in its effect."[4]

Complicating this predicament of advanced conventional weapons proliferation, we live in a time when "militarily relevant technologies" are becoming harder and harder to identify and classify. Technological advances, especially in the area of military systems, are a continuous, dynamic process; breakthroughs are always occurring, and their impact on military effectiveness and comparative advantage could be both significant and hard to predict at their nascent stages. Adding to this, many advanced *commercial* technologies, particularly in the field of information technologies, offer new and potentially significant opportunities for defense applications and, in turn, for increasing one's military power and advantage. Finally, such technologies and resulting capabilities rarely spread themselves evenly across geopolitical lines. In the case of the Asia-Pacific, for example, the diffusion of new and potentially powerful militarily relevant technologies—as well as the ability of militaries to exploit potential—varies widely across the region. This unequal distribution will, in turn, naturally affect how these technologies and capabilities may impact regional security and stability. Consequently, it is critical to assess the relative abilities of regional militaries to access and leverage new and emerging critical technologies (ECTs), their likely progress in doing so, and the impediments they may face, ultimately with an eye toward how it will affect relative gains and losses in regional military capabilities.

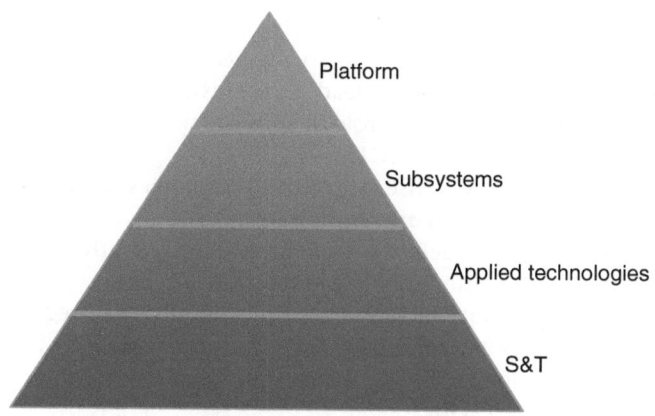

Figure I.1   The hierarchy of military technologies

This volume of essays, therefore, addresses the potential impact of ECTs on military capabilities—and therefore military advantage—in the Asia-Pacific region. By ECTs, we mean new types of research and development (R&D) breakthroughs, know-how, and artifacts that could have a significant, altering effect on how militaries fight and gain superiority over their rivals or potential adversaries. In this regard as well, ECTs are defined as applied military technologies below the level of platforms and subsystems, and also above the level of basic science and technology (S&T) (see Figure I.1). In other words, the conceptualization of an ECT does not apply to weapons platforms such as fighter aircraft, submarines, or armored vehicles, nor to components such as radar and jet engines. Rather, it addresses new technological innovations that are increasingly moving out of the developmental stage and into platforms and subsystems, and how these innovations might create new advantages for local militaries.

In addressing the issue of ECTs and their potential impact on military capabilities and advantage in the Asia-Pacific, it is important to address the following questions:

- How does one define a "militarily crucial" or "militarily relevant" ECT? Which kinds of ECTs are *most likely* to have an impact on the balance of power in the Asia-Pacific?
- How might one measure the impact of such ECTs on balances of power, particularly in terms of creating comparative advantages for a nation's military over a potential rival?

- What factors are more likely to drive technological advance in the Asia-Pacific? Does possessing a strong S&T base or having access to such technologies naturally lead to possessing a military advantage (exploitation)? What is the role of spin-on/spin-off when it comes to creating militarily useful ECTs?
- How do national defense acquisition strategies affect ECT-related technology innovation and exploitation?
- How do national defense industrial policies affect ECT-related technology innovation and exploitation?
- How does the globalization of S&T—especially with regard to commercial dual-use technologies, such as information and communications technologies (ICTs)—affect the diffusion of military-relevant ECTs to the Asia-Pacific?
- How might the unequal distribution of such ECTs affect military capabilities in the Asia-Pacific and, therefore, balances of power? Are less technologically advanced militaries doomed to inferiority? Or are there "offsetting" alternatives that a country may pursue in order to asymmetrically compete with a more technologically advanced rival?

Obviously, when it comes to their impact on Asian-Pacific military capabilities—and, therefore, their potential for creating significant military advantage—not all ECTs are created equal. Some ECTs are much more likely to have a disruptive effect on regional military balances. For example, *cyber*—that is, the ability to carry out computer-network attacks, hacking, cyber-espionage, and so on, together with the attendant defenses *against* such operations—is probably one of the more potentially disruptive ECTs, in that (1) it is less an "enabling" technology (i.e., something that is "nice to have") and more a "game-changer" technology that can have a fundamental impact on military capabilities; (2) it is one of the most easily proliferative ECTs; and (3) it has a high cost-benefit potential (i.e., relatively inexpensive to acquire and use offensively, while at the same time being relatively expensive to defend against).

This volume is organized into three parts. The first five chapters deal with the broader strategic issues relating to ECTs: what do we mean by ECTs; what are their potential impact on military capabilities; how might nations exploit ECTs for military advantage; and how such technologies are affecting (or could potentially affect) military modernization in the Asia-Pacific. The second part is a series of case studies of possible ECTs and how they could affect military capabilities, specifically (1) cyber; (2) air-independent propulsion for conventionally powered submarines; (3) supersonic and hypersonic propulsion for cruise missiles; and (4) automotives. Finally, the book concludes with three chapters that discuss the challenges facing countries—in both the Asia-Pacific and elsewhere around the world—when it comes to exploiting the promised potential of ECT, in particular, how technology can frequently over-promise and under-deliver; how often people make failed predictions and experience unanticipated outcomes with military innovation; and how radical innovation and the exploitation of ECTs are frequently less necessary to crafting military advantage than other, less dramatic forms of innovation.

Notes

1. Keith Krause, *Arms and the State: Patterns of Military Production and Trade* (Cambridge: Cambridge University Press, 1992), p. 19.
2. Krause, *Arms and the State*, pp. 18–26; Robert Gilpin, *War and Change in World Politics* (Cambridge: Cambridge University Press, 1981), pp. 175–185.
3. Richard A. Bitzinger, "Come the Revolution: Transforming the Asia-Pacific's Militaries," *Naval War College Review*, Autumn 2005, vol. 58, no. 4, pp. 5–6.
4. Bitzinger, "Come the Revolution," p. 6.

1

Emerging Technologies and Military Capability

Andrew D. James

"Emerging technologies" are the subject of considerable interest to academics and practitioners not only in the field of international security but also in the fields of economics and business. Emerging technologies are said to have the potential to change "the rules of the game" whether that "game" is the balance of military power between security actors or the balance of competitive advantage in a market between incumbent companies and new entrants.

By "emerging technologies," this chapter will mean new technologies that are at an early stage in their development. Their emergent nature means that they are characterized by considerable uncertainty: will their apparent technological promise be fulfilled? How long will it take to develop them to reach a sufficient state of maturity that they have practical application (and how much will that cost?). How might they be most profitably utilized? Examples of the effects of the emergence of new technologies on business are many and varied. Take the dramatic fall of Eastman Kodak. The dominant company in the photographic industry for a century was swept away in a matter of a decade by the emergence of digital imaging technology and the capacity of new entrants to exploit that technology in new products. Emerging technologies have had similar impacts on military power. During the Second World War, the emergence of radar had a dramatic impact not least in the defense of the UK. during the Battle of Britain and the conduct of anti-submarine warfare in the North Atlantic.[1] During the Cold War, emerging computer technologies, electronic component technologies (not least semiconductors), and propulsion technologies—all sponsored at the time by the military—each had significant impacts on the performance of Cold War weapons systems and perceptions of the Cold War balance of military power.[2]

The aim of this chapter is to examine the nature of emerging technologies and their potential impact upon military capability. This chapter is structured as follows: The next section provides some examples of emerging technologies that have been identified as having potential implications for military capability in the future. This section also discusses why emerging technologies are of concern in the military context and the threats and opportunities that they can pose. The following section defines "emerging technologies" and makes the distinction between weapon systems, technologies, and innovation. The chapter then introduces the notion of the "technology life cycle" to explain the nature of emerging technologies before introducing a key feature of emerging technologies, namely uncertainty. This leads into a discussion of the reasons why it is difficult to make accurate ex ante assessments of the rate and timing of a technology's development. The chapter goes on to consider the link between emerging technologies and military capabilities and the importance of institutional factors and the acquisition system in determining the speed of adoption of emerging technologies. The final sections of the chapter consider the sources of emerging technologies of military relevance in a global technological environment characterized by "Joy's Law" (i.e. "No matter who you are, most of the smartest people work for someone else"), and some implications for the Asia-Pacific.

Emerging technologies and the military

Visions of the military future almost always have a strong technological element. A review of futures studies conducted by the likes of the UK Ministry of Defence Development, Concepts and Doctrine Centre (DCDC)'s Strategic Trends Programme, the US National Intelligence Council Global Trends Program, the French Ministry of Defence and the European Defence Agency shows that emerging technologies feature prominently.[3] Advances in microsystems, nanotechnology, unmanned systems, communications and sensors, digital technology, bio and material sciences, energy and power technologies, and neuro-technologies are all identified as likely to have important applications in the defense sector. Cyber security and cyber warfare will grow ever more significant. The UK MOD DCDC's analysis is typical:

> Trend analysis indicates that the most substantial technological developments are likely to be in the areas of: ICT, sensor/network technology, behavioural and cognitive science, biotechnology,

materials, and the production, storage and distribution of energy. Advances in nanotechnologies will underpin many breakthroughs. Developments in individual areas are likely to be evolutionary, but where disciplines interact, such as in the combination of cognitive science and ICT to produce advanced decision-support tools, developments may be revolutionary, resulting in the greatest opportunities for a novel or breakthrough application...[S]ome [emerging technologies] may have catastrophic effects or present potential threats, perhaps through perverse applications, such as the use of genetic engineering to produce designer bio-weapons. (pp. 135–136)

Emerging technologies matter to the military because new technologies can present a threat or opportunity and yet they are veiled in uncertainty. The military understands the potential of new technologies but—like its counterparts in civilian business strategy—the uncertainty that characterizes emerging technologies means that they cannot know which emerging technologies are likely to mature to have profound impacts, how long that maturation will take, nor the technological trajectory. Most emerging technologies represent incremental improvements to what went before and enhance the competencies of the military along dimensions that they have traditionally valued. This kind of technological development presents relatively few challenges to the military, although their insertion into existing platforms can be difficult (as we shall see). In contrast, it is new technologies that are radical, competence destroying, and create new sources of military advantage along dimensions not traditionally valued or poorly understood by the military that tend to be the focus of attention and concern.

Fundamentally, these types of new technologies can change the environment in which military forces operate. A radical new technology can change the balance of power or create new forms of insecurity. The most dramatic illustration of the impact of new technology was the Allied development of the atomic and hydrogen bombs during the Second World War and the subsequent development of similar capability by the Soviet Union. In turn, the development of inertial navigation technologies added the prospect of accuracy to devastating lethality.

New technologies can redefine the way that warfare is conducted or create new types of warfare. Technology and military doctrine are closely coupled and interdependent.[4] Blitzkrieg, the AirLand Battle and Carrier Strike are but three examples of how new technologies combined with organizational change led to new ways of warfare.[5] The Internet and its widespread application have created the possibility of a new form of warfare—cyber warfare—that was hardly imaginable 20 years ago.

Equally, the significance of an emerging technology also depends in part on whether it is competence enhancing or competence destroying. An emerging technology that undermines existing training, equipment, doctrine and so forth will have a more dramatic impact on the military than one that complements or enhances existing military competencies. New technologies can render existing defense systems obsolete. Cavalry on the Western Front is but one example (although it was only the carnage of battle that brought this home to military planners).

At the same time, a new technology can provide new and more effective military capability. Precision munitions, not least the use of GPS technology, is a good example. Increased accuracy has led to a reduction in the number of aircraft required to attack targets and the substitution of lighter fighter-bombers for heavy bombers.[6]

By and large, attention has tended to focus on radical new-to-the-world technologies yet novel combinations of existing and mature technologies can also have profound military implications. Schumpeterian thinking emphasizes that innovation can be new combinations of existing technologies and stresses the potential significance of combining existing technologies in a new use. The DCDC Strategic Trends study identifies the rapid asymmetric insertion and exploitation of commercial technologies as a significant concern. Indeed, the experience of Iraq and Afghanistan provides graphic illustrations of how such tactics can have devastating effects. The contrast between the rates of combinatorial innovation of this kind has posed challenges to the traditional defense acquisition process. In the future, such developments may present ever-greater challenges to the traditional, long-term requirement and acquisition cycles.[7]

Defining emerging technologies

Before going any further, it is important to define what is—and what is not—meant by "emerging technologies." The UK's Defence Technology Plan defines emerging technologies as follows: "Emerging technologies can be characterized as: immature technologies in the early proof-of-principle stages; more mature technologies but where a novel defense application has been identified." While this definition appears clear and straightforward (and this chapter will use it), it is the case that a feature of much of the discussion of emerging technologies is a lack of clarity as to the subject of analysis.

"Emerging" is used variously to examine technologies that analysts regard as potentially emerging in the far future (e.g., the latest UK MOD DCDC programme report looks out to 2040 and consciously examines what technological developments *may* occur). In contrast, "emerging" is sometimes used to describe technologies that have reached a stage that we know that they *will* find application in a weapon system in the near future (e.g., many of the "emerging" IT technologies discussed by Bruce Berkowitz in his 2003 book are now in military service, at least with the US military[8]). Sometimes analysts conflate the far future and the soon-to-be fielded as "emerging technologies" giving the impression to the unwary that (true) emerging technologies on the technological far horizon are as certain to be fielded as those in late stage development. This raises important questions about timing that are critical to discussions about emerging technologies. It also raises issues about uncertainty. Both issues will be discussed later in this chapter.

A further source of ambiguity in discussions about emerging technologies is what is meant by "technologies." Technologies can be defined as "The ensemble of theoretical and practical knowledge, know-how, skills and artefacts that are used...to develop, produce and deliver...products and services."[9] This definition is concerned with technology and business but it holds equally for military technology. Military technology combines "theoretical and practical knowledge"—some may be science based but much will be engineering knowledge, including "know-how and skills"—individual and collective knowledge that arises within defense through "learning by doing," team working, culture and so forth and "artefacts"—tangible assets such as capital equipment, manufacturing facilities and so forth. It is worth noting that following this definition much of the core "technology" that underpins defense is intangible and human.

There is an important distinction here that is sometimes missed by military analysts of emerging technologies (business analysts miss this too). The distinction is between technologies and products/services (in the case of the military, we mean weapons, their delivery systems and the infrastructure that supports military capability). Technologies underpin weapon systems but are distinct from them. Militaries want "capability," not technologies *per se*. Consequently, how emerging technologies and other factors are combined into military capability should be the critical consideration not the emerging technologies themselves (this is an important point that we shall return to later).

The decision to invest in an emerging technology in the hope of military capability advantage depends on very many factors, not least the perception of the threat environment. The Cold War was different to today. The military needs of forces in Iraq and now Afghanistan have brought home the fact that emerging technologies are only of military significance if they can be matured and fielded quickly enough to make a difference to current combat operations. Investments in emerging technologies that may only have application in 30 years' time and are characterized by uncertainty have always had lower priority. Constrained defense budgets in the UK, Europe and the United States mean that this is likely to be even more the case in the future. Indeed, this speaks to the need for greater agility in the defense acquisition process. The military technological innovation timescale that emerged during the Cold War means that development times of 20 years for major weapon systems became the norm. New designs of improvised explosive devices (IEDs) appeared in Afghanistan on a monthly basis. Changes in cyber threats can occur just as quickly. This requires reform of the defense innovation process to promote greater agility and reduce time-to-fielding of new equipment.

Another important point needs to be made and that is the danger of analysis of emerging technologies degenerating into some form of technological determinism. The idea that emergence of a new technology leads inevitably to change and that technology is necessary and sufficient to drive innovation in military capability has been widely discredited by those who study innovation. The study of military innovation emphasizes the critical role of political and bureaucratic politics among both military and civilian actors in selecting (or not selecting) particular technologies.[10] Equally, it emphasizes the relationship between technology and doctrine.[11] Grissom summarizes the literature on social shaping of technology and its emphasis on the nature of technologies as:

> ultimately ideas that are shaped by discourse and competition with different views on the potential of a given technology ... these interest groups (such as research teams, policymakers and investors) vie to superimpose their own vision on a developing technology by building a coalition around their vision, engaging in bureaucratic maneuvers to exclude other groups, and ensuring that important design and engineering choices reflect their vision for the technology.[12]

In short, an emerging technology, its funding, trajectory and adoption in use is shaped by a variety of actors. There is nothing "inevitable" about the trajectory of a new technology or how it will (or will not) be used. This insight is important as we turn to consider the technology life cycle.

Emerging technologies and the technology life cycle

A clear understanding of what we mean by "emerging technologies" matters since there is a danger that those discussing the military implications of such technologies may find that they are talking at cross-purposes about different objects of analysis, over different timescales and so forth. Those who study technological change think in terms of the technology life cycle (TLC).[13] This S-curve is illustrated in Figure 1.1.[14] Note that the TLC is divided into three stages distinguishing between emerging, transitional, and mature technologies and is mapped along two dimensions: time and performance. In many respects, this is covering similar ground to the idea of Technology Readiness Levels (TRLs) used by NASA, the European Space Agency and in defense.[15]

The TLC begins with the emerging phase. An emerging technology is characterized by its relatively poor and uncertain performance. The technology is at the proof of concept stage, characterized by high levels of technological uncertainty and uncertainty as to the feasibility of its application in military systems. The emerging stage may involve the transition from scientific research to applied research and the observation of the essential characteristics of the technology. Analysis and experimentation will likely take place to ensure proof of concept. At the emerging stage, the technology is a long way from providing military

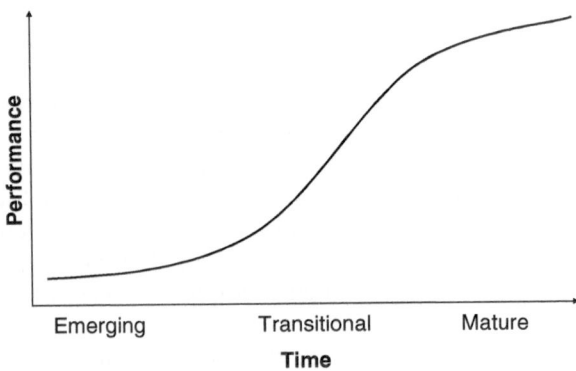

Figure 1.1   Emerging technologies and the technology life cycle

capability in a fielded system (TRL 1–3). During the transitional stage (roughly TRL 4–5) the technology is subject to testing through prototyping and other activities. At this stage, the technology will likely be tested in a relevant and realistic environment to judge its potential performance. At the mature stage, the technology's performance characteristics are well understood. The technical uncertainties that characterized the earlier stages have been reduced and efforts focus on questions of design and integration into military systems prior to the production phase for a new military system or the insertion of the new technology into an upgrade of an existing system.

The issues of what we mean by "time" and "performance" are worthy of further examination. The TLC has time as its X-axis and it is clear that the timescale for a new technology can vary greatly depending on its technical characteristics, complexity, the state of scientific and engineering knowledge as well as level of funding available for that technology and the priority it is given within the defense acquisition community of an individual country. The Y-axis of the TLC is performance. This should also be examined carefully although the unit of performance is frequently left unexplained by academics using the TLC approach. In computing, performance may be memory size or clock speed. In the military context, performance may be speed, lethality, or precision or perhaps some combination of performance measures. In the modern security environment, what constitutes the key performance measure is increasingly open to debate and no longer straightforward. Performance is no longer about only technological trajectories but also about whether technologically possible weapons are suitable on political and ethical grounds. Budgetary austerity in Europe and the United States makes the cost effectiveness of new technologies an important consideration.

Uncertainty and emerging technologies

The uncertainty that surrounds emerging technologies has been mentioned at various points in this chapter and deserves further discussion. Uncertainty is a key characteristic of technological change and stems from the difficulties of ex ante assessment of the rate and timing of a technology's development. Failed predictions about technological developments are legion. Bill Gates is reputed to have said about computer memory that "640k ought to be enough for anybody."[16] A British military training manual in 1907 stated: "It must be accepted as a principle that the rifle, effective as it is, cannot replace the effect produced

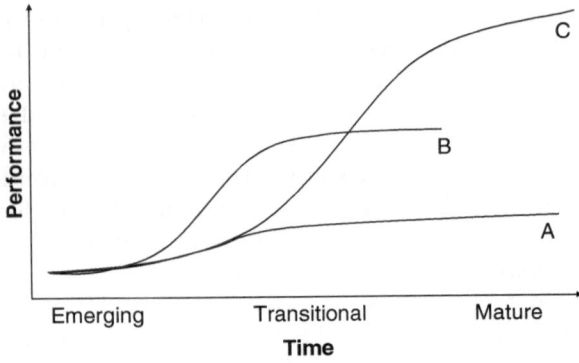

Figure 1.2   Emerging technologies and the nature of technological innovation

by the speed of the horse, the magnetism of the charge and the terror of cold steel." Marshal Ferdinand Foch was reported as saying in 1911: "Airplanes are interesting toys but of no military value."[17]

Why are emerging technologies characterized by uncertainty? Figure 1.2 describes three S-curves for the life cycles of three technologies. They illustrate that uncertainty may arise because the technology is radical; the technology is early in its life cycle but also because of the increasingly amorphous nature of security threats.

The S-curve labeled (A) emphasizes that some technologies may fail to deliver on the early performance claims of their advocates. They may fail to deliver because engineering and technological challenges arise that are difficult to overcome without excessive time or other resources. Equally, they may reflect a "conspiracy of optimism" in which those who have a vested interest in a particular technology over-sell its military potential. Academic grant holders and researchers in government defense research laboratories may boost the technology to ensure their own funding. Entrepreneurs may claim military relevance to access "free" defense funding for early stage technology development. Experts in think tanks may over-sell the potential of a technology to sell books, secure funding for workshops or get access to decision makers. Failure is normal, natural, and desirable. This may be an uncomfortable truth for officials in a resource-constrained environment. Historically, the Defense Advanced Research Projects Agency's (DARPA) "success" has been its ability to allow space for "failure"—a number of significant DARPA programs (including UAVs) were the outcome of returning to a "failed" project.

The S-curve labeled (B) illustrates a situation where a technology is superseded by other technologies that are better or cheaper or faster to develop. Since it is impossible to forecast the eventual outcomes of emerging technologies, and since few (if any) countries have the resources to pursue all emerging technologies, this raises important strategic questions: should a country seek to pick a winner? Should it lag behind and hope that an ally invests in the technology, and is subsequently willing to transfer that technology, or does a country seek to follow all of them but develop processes that allow it to know when to stop when initial expectations prove unfounded? A small country may choose to lag behind and simply invest in absorptive capacity (to access commercial technologies) and/or a limited range of distinctive capabilities that allows it to enter into cooperative arrangements with other countries. S-curve (C) emphasizes that only some technologies will reach the stage that they are deemed sufficiently mature that they may be considered for transition into new military capability. The design dilemma faced by the defense acquisition community is the trade-off between the costs of designing-in emerging technologies against the benefits. Advocates of the emerging technologies understandably focus on their benefits but the costs are non-trivial. Those costs include those related to the uncertainties surrounding any emerging technology (will it work, at what cost and when?); the disruption to established military ways of operating (including doctrine); and the costs of substituting the new weapons for existing weapons. The fielding of new technologies in weapons systems is a function of the weapons development process and the procurement process of individual governments. An emerging technology may move along a development trajectory but never be translated into a fielded weapon.

Emerging technologies and military capability

There is an understandable tendency for those who analyze emerging technologies to focus on those technologies. However, military innovation is about more than emerging technologies and there is a long road from the emergence of a technology to its having an impact on military capability.

An important point here is that the knowledge that underpins these emerging technologies rarely resides in the military of a single country. The United Kingdom was not the only country to possess the knowledge that underpinned the emergence of radar. The United States was not the only country during the Cold War to possess semiconductor technologies. The military-scientific-industrial establishments of each country made decisions to prioritize the development of certain technologies for certain applications. The pace of application varied in part as a consequence of the ability of the military procurement process to pull the technologies through into fielded weapons.

In short, military innovation is about more than an emerging technology. The idea that military capability can be reduced to the fielding of weapon systems with superior technology is plainly wrong (although in the eyes of much of the world this Cold War mindset has characterized aspects of US thinking with disastrous consequences, not least in Afghanistan). History shows that improving military effectiveness may require complementary organizational and doctrinal innovation.[18] Indeed, it has been argued that the whole issue of emerging technologies would be better thought of in terms of the innovative and emerging "uses" for existing technologies (it will be recalled that this point is made in the UK MOD's Defence Technology Plan). Recent insurgencies prove that a 100-year-old rifle is as effective as a modern equivalent when used effectively and IED designs do not need to evolve that much to remain devastatingly effective.[19]

The procurement process can also have an important part to play in the adoption of an emerging technology. The responsiveness of the procurement process to new technologies has always been an important concern and is becoming more so in the face of the twin forces of budget constraints (at least in Europe and the United States) and rapid technological change. The defense acquisition process has always faced the challenge that it generates more ideas and potential new technologies than it can possibly transfer into new weapons and other military equipment. This raises the matter of the so-called "Valley of Death"— the situation where R&D supports the development of an emerging technology only for that new technology not to transition into fielded equipment due a lack of procurement funding. Even where defense spending is growing as it is in the Asia-Pacific, it will likely remain the case that the number of potential technological ideas will exceed opportunities to transition those into equipment program.

Rapid technological change also poses a challenge for the procurement process. Technology cycles in the civilian economy are shortening (witness the rate of change in integrated circuits) but weapons systems take longer to develop. This makes it increasingly difficult to take advantage of commercial innovations. "New" weapons are introduced that include technologies that are already obsolete. This raises huge questions for the defense acquisition process. How to make the defense acquisition process more agile and responsive to emerging technologies? How to insert a new technology when it is "good enough"? Various responses have been adopted or proposed including open system architectures and "spiral development" and yet the divergence of the technology life cycles between civilian and defense equipment remains a profound challenge for the military acquisition process.

Critically, the adoption of a new technology is dependent on the response of the military. In the civilian economy, competition stimulates innovation and determines which new technologies are adopted. In the military innovation system, which new technologies are adopted is determined by bureaucratic and political decisions. In peacetime, the military is characterized by conservatism towards innovation. Military innovation arises out of inter- and intra-service rivalries as well as interactions and negotiations between the military and civilian "champions" of innovation.[20]

Joy's Law, "military" technologies and open innovation

A further point needs to be appreciated if we are to understand the nature of emerging technologies. We no longer live in the 1960s when US defense R&D spending accounted for something like half of all defense *and* non-defense R&D spending in the world. At that time, US defense R&D and procurement was able to stimulate whole new technologies (like semiconductors). However, this state of affairs did not last. Defense R&D remains important for the development of certain defense-specific technologies and knowledge but, more broadly, the defense innovation system has grown isolated from civil technology developments as a consequence of economic and technological change. Many (most) of the emerging technologies identified as potentially important to defense are not of defense origin but are emerging from commercial R&D activity taking place in civilian sectors, SMEs and start-ups and in universities throughout the world.

The changing dynamics of technology mean that defense in the United States and Europe is increasingly having to recognize and accept Joy's Law. As Bill Joy, Co-Founder of Sun Microsystems, said: "No matter who you are, most of the smartest people work for someone else." Joy's Law represents a profound shock for the defense innovation systems of many countries, not least the United States (where, for a long time, many of the smartest people did work in US defense).

The implications of Joy's Law have already been accepted in many sectors of the civilian economy where companies increasingly practice what has become known as "open innovation."[21] Open innovation is the idea that organizations should seek, engage and exploit knowledge wherever it resides.[22] Under open innovation the ability to engage effectively with external suppliers of technological knowledge becomes critical to the innovation process.

To take advantage of open innovation, the defense innovation process will need to place more emphasis on the efficient exploitation of technological knowledge wherever it resides and take advantage of the significantly greater investments made in markets outside of its control or influence. The emphasis will have to be on exploiting and integrating technologies rather than large investment in new cutting-edge technologies in all but a few defense critical areas. Accordingly, one of the key competencies that defense will need to develop is that of *absorptive capacity*—the ability to be able to recognize valuable external knowledge, assimilate it, and apply it.[23] Absorptive capacity in the defense context will mean the ability to (i) develop effective search mechanisms to identify potentially important external technologies outside the traditional defense innovation system(ii) build effective partnerships with (potentially) non-traditional suppliers of such technological capabilities, and (iii) find means to exploit those capabilities to military advantage.

Policy implications

A number of policy implications arise from this discussion. First, emerging technologies can have significant implications for military capability but the path from technological emergence to military capability is a long and uncertain one. Many immature technologies fail to live up to the promises of their advocates. Hype is common; failure even more so. The nature of the acquisition process not least its agility and responsiveness to new technologies is critically important. Equally so is the recognition that the combination of mature technologies in use can also have profound implications for military capability. The shock of the old can be just as great as the shock of the new.

This raises a second point, namely whether we should use an absolute or relative measure in judging whether a technology is "emerging." What is a mature technology in one country may be an "emerging" technology for another country or region (in our case the Asia-Pacific). This raises important issues about the diffusion of military technologies and innovation that were the subject of an excellent collection edited by Emily Goldman and Leslie Eliason.[24] Arms transfers and cooperation play an important role in this process and will grow as European,

US and other governments and companies seek to gain a share of growing defense procurement budgets in the Asia-Pacific. Equally, the relative capabilities of national innovation systems are likely to be an important consideration. Most of the true emerging technologies identified in futures studies are emerging globally, and defense is likely to play only a minor role as sponsor and user. The strength of national innovation systems (rather than just defense innovation systems) will be important. Cold War thinking about the dominant global position of the United States through its homegrown defense technologies is declining in relevance by the day.[25] The further growth of the already considerable scientific and technological capabilities of the Asia-Pacific region is likely to have significant implications in the field of emerging technologies.

The gulf between the most advanced technologies being applied for military use by the United States and those of much of the Asia-Pacific region is considerable. The difficulties (and cost) of trying to close the gap are enormous and beyond the scope (and ambition) of most regional actors. Since it is impossible to forecast the eventual outcomes of emerging technologies, and since few (if any) countries have the resources to pursue all emerging technologies, this brings us back to the strategic questions raised earlier in this chapter, namely: should a country seek to pick a winner, hould it lag behind and hope that an ally invests in the technology, and is subsequently willing to transfer that technology, or does a country seek to follow all of them but develop processes that allow it to know when to stop when initial expectations prove unfounded? The use of more mature technologies in new ways is a more likely direction of development for Asia-Pacific militaries. We are already seeing such developments, not least in the development of military capabilities in cyberspace.

Conclusion

The aim of this chapter has been to examine the nature of emerging technologies and their potential impact upon military capability. This chapter has defined "emerging technologies" as new technologies that are at an early stage in their development or relatively mature technologies combined in new ways. This chapter has examined the nature of emerging technologies, their implications for military capability, and the challenges that they pose to the acquisition system. The chapter has emphasized that their emergent nature means that they are characterized by considerable uncertainty: at its core, the chapter stresses that it is a potentially long and uncertain journey from the emergence of a new technology to its use in a fielded weapons system.

Such issues are important because new technologies have the potential to change the environment in which militaries operate and a radical new technology can change the balance of power or create new forms of insecurity. New technologies can change military doctrine and the way that war fighting is conducted. New technologies can make existing defense systems obsolete or provide new and more effective military capability. By and large, attention has tended to focus on new-to-the-world technologies yet—as this chapter has emphasized—novel combinations of existing and mature technologies can also have profound military implications.

How emerging technologies of military importance are identified and integrated into weapons systems through the acquisition process is a critical issue. The rate of technological change—driven by technological and economic factors mainly from the civil sector—places a premium on agility and responsiveness of the defense acquisition system. At the same time, it suggests the need for a profound shift from a closed towards an open model of defense innovation.

Notes

1. On radar and the air defense of the UK see the excellent PhD thesis by Phillip Judkins (2007) *Making Vision into Power: Britain's Acquisition of the World's First Radar-based Integrated Air Defence System 1935–1941*, PhD thesis, Defence College of Management and Technology, Cranfield University.
2. For more details, see Andrew D. James (2007) "Science and technology policy and international security," in Brian Rappert (ed.) *Technology and Security: Governing Threats in the New Millennium*, Palgrave MacMillan: New York and Houndsmill.
3. James, A.D. and Teichler, T. (2014) "Defense and security: new issues and impacts," *Foresight*. 16 (2): 165–175
4. Alic, J.A. (2007) *Trillions for Military Technology: How the Pentagon Innovates and Why it Costs So Much*, Palgrave Macmillan: New York and Houndsmill.
5. See Murray, Williamson R., and Millett, Alan R. (eds) (1996) *Military Innovation in the Interwar Period*, Cambridge University Press: Cambridge.
6. Alic (2007) *op cit.*
7. Development, Concepts and Doctrine Centre (DCDC) [UK], *Global Strategic Trends: Out to 2040*, Fourth Edition, Ministry of Defence: London.
8. Berkowitz, B. (2003) *The New Face of War: How War Will be Fought in the 21st Century*, The Free Press: New York.
9. Burgelman, R.A. and Rosenbloom, R.S. (1989) "Technology strategy: an evolutionary process perspective," *Research on Technological Innovation, Management and Policy*, 4: 1–23.

10. Grissom, A. (2006) "The future of military innovation studies," *Journal of Strategic Studies*, 29 (5): 905–934.
11. Alic (2007) *op cit.*
12. Grissom (2006) *op cit.*
13. The analogy of biological life is popular in the academic business and innovation community and the technology life cycle (and product life cycle which I will also mention) should not be confused with the R&D lifecycle, product life cycle management and so forth.
14. Technology life cycle and product life cycle are both presented as logistic (S) curves.
15. For a definition of Technology Readiness Levels see http://esto.nasa.gov/files/trl_definitions.pdf.
16. He says he never said this.
17. Simons, A. "The art of being wrong: failed predictions," http://urbantimes.co/2012/02/being-wrong-failed-predictions/ (Retrieved 23 January 2013).
18. Murray and Millett (1996) *op cit.*
19. I want to acknowledge with thanks the thoughts of Andrew Burton on this point.
20. See Grissom (2006) *op cit* for a review of the work of Posen; Rosen; Murray and Millet; Pierce and others.
21. Huston, L. and Sakkab, N. (2006) "Connect and develop: inside Procter & Gamble's new model for innovation," *Harvard Business Review* (March): 58–66.
22. Chesbrough, H. (2003) *Open Innovation,* Harvard Business School Press: Boston.
23. Cohen, W.M. and Levinthal, D.A. (1990) "Absorptive capacity: a new perspective on learning and innovation," *Administrative Science Quarterly*, 35 (1): 128–152.
24. Goldman, E.O. and Eliason, L.C. (eds) (2003) *The Diffusion of Military Technology and Ideas*, Stanford University Press: Stanford.
25. This is recognized if not always by some within the defense-industrial-scientific complex. For example, Committee on Science, Security, Prosperity, Committee on Scientific Communication, & National Security. (2009). Beyond'Fortress America': National Security Controls on Science and Technology in a Globalized World. National Academy Press: Washington DC.

2

The Potential Import of New, Emerging, and Over-the-Horizon Technologies

Andrew L. Ross

For both practitioners and analysts, new and emerging technologies have long been a central focus of attention. National security and defense planners and analysts attempt to identify and anticipate the impact of new and emerging technologies. They are a prominent feature of intelligence assessments, department or ministry white papers, defense planning exercises, military exercises and war games, and service wish lists. In the United States, for instance, concerns about new and emerging technologies, particularly their diffusion, pervade the series of White House national security strategies, OSD defense strategies, JCS joint visions and military strategies, joint and service doctrinal and operational planning documents, intelligence community forecasts, and think tank reports. Science and technology (S&T) and research and development (R&D) programs are expected to yield exploitable, and deployable, new technologies. Not so long ago, a much-vaunted transformation enterprise was to exploit the opportunities thought to be inherent in an information-technology revolution to bring about a game-changing information-technology revolution in military affairs (IT RMA). The National Intelligence Council's recent *Global Trends 2030: Alternative Worlds*, as *Global Trends 2025*, *Global Trends 2020*, and *Global Trends 2015* before it, highlights technological trends, challenges, and opportunities.[1] As if in search of a worthy successor to the Soviet threat of old, OSD's Office of Net Assessment and the U.S. Navy,[2] along with a motley crew of fellow travelers, hype Chinese military-technological and defense-industrial innovation.[3] In what must now qualify as an annual rite, analysts of varying persuasions and capabilities rush to judge whether the characterization of China's R&D and procurement programs provided in the Pentagon's annual report on *Chinese Military Power* is too hot, too cold, or just right.[4]

New and emerging technologies are a moving target.[5] Technology is not static but dynamic.[6] The extent and rate of change may vary considerably over time and across sectors and operational domains, but technological change is, essentially, constant. Technological change ranges across materials,[7] system components (including platforms and platform components), subsystems, systems and systems-of-systems; nodes, networks and networks-of-networks; hardware and software. It may be sustaining or disruptive, mundane or profound. Spin-on as well as spin-off is at work. A variety of change agents, with a range of capabilities, are at play: individuals (civilian and military), private and public sector organizations (firms, R&D enterprises, NGOs), violent non-state actors,[8] even international organizations or institutions,[9] as well as governments or states. And technology, even military technology and practice, spreads.[10] Global science networks and technology markets have undermined the efficacy of neo-mercantilist practices. Diffusion happens;[11] it has proven difficult to prevent, control, manage, or limit. Technological leaders cannot long expect to monopolize their innovations and technological latecomers, or followers, have been known to surpass the leaders.[12]

The development, production, procurement, deployment, and employment of new military technologies are of no small import. New and emerging technologies, and the expectations associated with them, have been central, even critical, to national security policy, strategy, force planning (whether military modernization or force structure, mix and size), and investment and spending debates and decisions.[13] At times the promise of a technological silver bullet has served to privilege particular capabilities (and services)—air, or aerospace, over ground, nuclear over conventional, for instance. The potential impact of new technologies may not be merely tactical or operational but strategic. They can right military imbalances or contribute to the creation of new imbalances. They may serve as a force multipliers, providing a qualitative edge thought to compensate for quantitative inferiority. Technological developments can reshape, even transform, offensive, defensive, and deterrent capabilities. The interface among offense, and deterrence may be altered, perhaps in unexpected and even unintended ways. Valuable new capabilities may be created (power projection, standoff attack, low-observability, anti-access, area denial, C4ISR networks); the value of existing capabilities tends to be undermined. The battlespace may be altered, even transformed. New arenas, or domains (i.e., cyber), may be opened to military competition. Arms races and security dilemmas can be stabilized or exacerbated. Battles and wars are won—or lost. The role of technology in how states prepare for and conduct war—in how militaries prepare for and deter, fight, and win wars—is ubiquitous.

New and emerging technologies

The classification of technologies as new and/or emerging cannot but be highly context dependent. Not only temporal but spatial considerations apply. What was new or emerging yesterday may not seem so novel today. Extant or even oft exercised capabilities for states in one part of the world may be new or emerging for states elsewhere (even those that have been, or fear that they could in the future be, on the receiving end of those capabilities). Technologies appropriately characterized as new or emerging[14] today include the following:

- Information technologies
  - New "big data" storage and processing solutions[15]
  - Data base-processing fusion
  - Virtual reality (VR)[16]

- Cyber
  - Offense
  - Defense

- Robotics (aka unmanned, or "uninhabited," systems)[17]
  - Air/aerospace
  - Surface
    - Ground
    - Naval
  - Subsurface
  - Micro[18]
  - Autonomous[19]

- Precision strike[20]
- Directed energy[21]
  - Offensive
  - Defensive[22]

- Low observable (LO) technology (aka stealth)

  - Active
  - Passive

- High-speed air-breathing systems

  - Supersonic
  - Hypersonic

- Conventional prompt global strike[23]
- Air-independent propulsion (AIP) for subsurface vessels
- C4ISR:[24] a network of networks or system of systems.
- Ballistic missile defense[25]
- Battlespace situational awareness
- Biomedical/mechanical engineering[26]

  - Human augmentation[27]
  - Implants

    - Retinal implants to enable night vision, for instance

  - Prosthetics
  - Bioelectronics[28]
  - Brain-machine interface (BMI)[29]
  - Powered exoskeletons
  - Psychostimulants
  - Neuro-enhancements
  - Neuro-pharmaceuticals
  - Augmented reality systems[30]

- Non-, or less-than-, lethal technologies

  - Counterpersonnel
  - Countermaterial

- Digital manufacturing

  - 3D printing

    - Additive manufacturing[31]

As they seek to maintain or enhance their lead, dissuade competitors, catch up with or surpass a leader, keep pace, or merely avoid falling further and further behind, leading and trailing militaries alike devote not inconsiderable resources to S&T programs that may yield new and emerging technologies such as these and to R&D programs that yield new capabilities to be procured and deployed.[32]

Over-the-horizon "technologies"

Today's over-the-horizon "technologies" will be tomorrow's new or emerging technologies. They will emerge from scientific inquiry that is distinctly transgressive, that flagrantly trespasses across the disciplinary boundaries still regarded sacrosanct by academic researchers oblivious to the emerging science landscape.[33] Emerging disciplinary fusion will yield basic and applied scientific research that is not merely interdisciplinary but multi- and transdisciplinary (and produced, as often as not, by networked, multi-institutional, international teams[34]).[35] Likely suspects, all already at least interdisciplinary, for the taboo-busting research that will yield potential future game changers include:

- Genomics/genetic engineering[36]
- Mathematical/computational biology
- Synthetic biology[37]
- Biomedical engineering[38]
- Cognitive/neuroscience[39]
- Artificial intelligence[40]
- Quantum computing[41]
- Biological, or bio/neuro-inspired, computing (aka neurocomputation)
- .Nanotechnology
- Nanoscience
- Materials science[42]
- Geo-/environmental engineering[43]
- Science of complex systems/chaos and complexity theory[44]
- Forecasting

Coming to a future battlespace near you: New life forms—microorganisms, perhaps—that are the result of "human-directed evolution." Networked, genetically enhanced, computationally and/or mechanically modified warriors with seamless human-machine interface—yes, cyborgs (you may not know it, but you have already been assimilated and you like it)—and exquisite shared situational awareness (facilitated by an abundance of micro- and nano-sensors that result from leap-ahead advances in material sciences). Not only hardware but warriors—human nodes—will plug into the networks of networks, and systems of systems, and "play." Fast, powerful, secure computing based on qubits rather than bits. Virtual operations that that yield kinetic effects. Integrated manned and unmanned platforms—directed and autonomous, micro as well as nano—and systems able to exploit quantum communications and teleportation capabilities[45] within and across operational domains. The ability to achieve spatio temporal cloaking by creating and exploiting a "time hole."[46] Tactical, operational, and strategic—preemptive, even preventative—manipulation of physical and virtual environments. These and other as yet unanticipated capabilities can be expected to emerge from today and tomorrow's basic research. The confluence of such an imposing array of potentially disruptive technologies and game changers could amount to transformation on steroids.

Technology and innovation

The extent to which new and emerging and over-the-horizon technologies will be disruptive or game changing depends on how effectively they will be utilized. That will be determined by organizational and doctrinal developments. Technology is but one of the three components of military innovation.[47] For the potential of new and emerging technologies to be fully exploited or realized in Asia-Pacific or elsewhere, particularly if not only their operational but strategic potential is to be realized, technological change must be accompanied by complementary organizational and doctrinal accommodation (at the least) or change. These three components of innovation—technology, organization, and doctrine (or operational art)—constitute the innovation triad (Figure 2.1). The triad captures "hardware" and "software" and product and process innovation. It is technology, of course, that provides the hardware dimension of innovation and its concrete products.[48] Organizational and doctrinal changes, the software of military innovation, feature what is characterized in the broader literature as process innovation.[49]

Technology is the most visible and concrete component, or manifestation, of military innovation. For both the military and the analytical community, the allure of new technology has proven difficult to resist. However, there is more to military innovation than new, even breakthrough, technology. Militaries must learn how to effectively utilize new technologies.[50] If new technology or hardware is to be exploited effectively, organizational adaptation (at a minimum) or, even, recreation[51] and/or doctrinal refinement or reformulation is typically required.[52] Incremental technological advances may mesh well

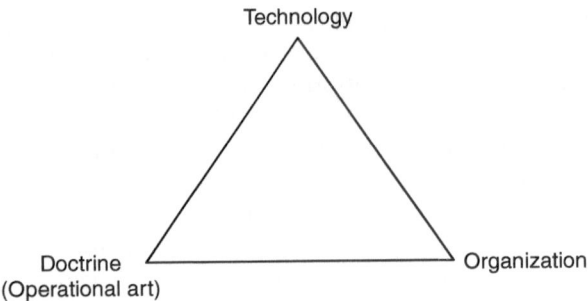

Figure 2.1   The military innovation triad

with existing organizational structures and capabilities, but new technologies can encounter organizational and bureaucratic resistance.[53] It may be necessary to restructure organizations or even develop new organizations with new skill sets. Similarly, while existing doctrine may accommodate sustaining technological advances, new and emerging technologies, particularly breakthrough technologies, may well require rethinking the principles that inform the employment of military force.

The components of military innovation rarely change simultaneously or at the same rate. Technology, for instance, may leap ahead, requiring organizational and doctrinal catch up. Doctrinal or operational vision can drive organizational change and drive technological development. How hardware and software, and product and process, innovation, are integrated (or are not) is of tremendous import. It determines the extent of change, whether it will be modest or profound, evolutionary or revolutionary.

Most military innovation is incremental. Existing capabilities are improved upon. Indeed, "modernization" consists of essentially incremental advances that qualify as sustaining innovation—the routine improvement of existing capabilities in support of existing performance metrics.[54] The discontinuous, disruptive innovation that that underlies military revolutions and transformations is extraordinary rather than routine. Technology is not merely improved, it leaps ahead. Doctrine is not just refined, it is reformulated. Organizations do not only evolve, they are (re)created anew.

What distinguishes military revolutions, revolutions in military affairs (or military-technological revolutions), or military transformations is the combination of novel (1) technology (new weapons and new weapons systems), (2) doctrine/operational art (new ways of fighting), and (3) organizational structures or institutions. It is the *confluence* of

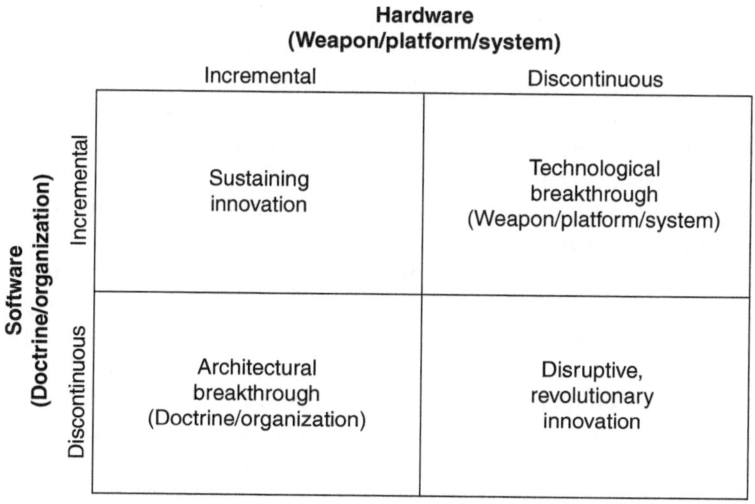

Hardware
(Weapon/platform/system)

	Incremental	Discontinuous
Incremental	Sustaining innovation	Technological breakthrough (Weapon/platform/system)
Discontinuous	Architectural breakthrough (Doctrine/organization)	Disruptive, revolutionary innovation

(Left axis label: **Software (Doctrine/organization)**, with subdivisions Incremental / Discontinuous)

Figure 2.2   Military Innovation Matrix

these three components of innovation that can yield discontinuous, radical breakthroughs.[55] Despite the technologies featured in many of the labels employed to characterize military revolutions or revolutions in military affairs,[56] technological change alone does not make for a revolution, or transformation. Operational and organizational change is also required.

This confluence is not a frequent occurrence. Despite the attention they have attracted, revolutions are quite rare. Those identified by Krepinevich and Murray and Knox are, essentially, innovation outliers.[57] Military innovation is most frequently found in quadrants of the military innovation matrix (Figure 2.2) other than that within which disruptive, revolutionary innovation appears.

Discontinuous innovation that poses the potential of technological or architectural "breakthroughs" is uncommon and, typically, unanticipated. As Stefik and Stefik note, breakthroughs "create something new or satisfy a previously undiscovered need"[58] and enable militaries to do something that wasn't thought possible. Their uses and consequences may be unintended. Existing practices may be transformed or displaced.

As depicted in the matrix, discontinuous technological innovation, even in the context of sustaining doctrinal and/or organizational change, constitutes a "technological breakthrough." Discontinuous doctrinal and/or organizational—or software—innovations are architectural breakthroughs. Architectural innovation redefines how the components of technologies, doctrines, or organizations are linked and features major changes in the relationships among and integration of technology (hardware) and doctrine and organization (software).[59] Technology used in ways it hasn't been used before as result of dramatic departures in operational concepts or organizational structure qualifies as an architectural breakthrough. Discontinuous technological and architectural innovations, or breakthroughs, both occur much less frequently than sustaining innovation.

Disruptive, revolutionary innovation occurs when discontinuous technological, doctrinal, and organizational changes effectively converge; discontinuous hardware and architectural changes come together in a coherent, integrated whole, even though they may well not have occurred simultaneously.[60] Existing capabilities are rendered obsolete and displaced, not merely optimized. New dominant technologies, doctrines, and organizations are developed and integrated as never before. New performance metrics are adopted.[61]

The potential and import of new and emerging technologies—hardware—depends on whether they represent sustaining or disruptive, routine or profound innovation. And both today's new and emerging technologies and over-the-horizon technologies must be accompanied by requisite organizational and doctrinal (or operational)—software—developments if their potential impact on regional and international security affairs is to be fully realized (particularly if that impact is to be strategic rather than merely operational). Innovation capacity is determined as much by architectural "software" capabilities as by technological hardware capabilities.

Notes

1. See http://www.dni.gov/index.php/about/organization/national-intelligence-council-global-trends, accessed 26 December 2012. See also the U.S. Army's "Unified Quest" futures study program; info @ http://www.arcic.army.mil/unified-quest.html, accessed 30 December 2012.
2. Joined of late, with Air-Sea Battle, by the USAF. See Richard Bitzinger, "AirSea Battle: Old Wine in New Bottles," RSIS Commentaries, No. 159/2012, 23 August 2012, available @ http://www.rsis.edu.sg/publications/Perspective/RSIS1592012.pdf?utm_source=getresponse&utm_medium=email&utm_campaign=rsis_publications&utm_content=RSIS+Commentary+159%2F2012+AirSea+Battle%3A+Old+Wine+in+New+Bottles%3F+by+Richard+A.+Bitzinger, accessed 30 December 2012.
3. The Soviet bear's successor, apparently, is the Chinese dragon. Really? A mythical creature? Is that the best American threat mongers can do these days?

4. Office of the Secretary of Defense, *Annual Report to Congress: Military and Security Developments Involving the People's Republic of China*, Washington, DC: Department of Defense, 2014, available @ http://www.defense.gov/pubs/2014_DoD_China_Report.pdf; accessed 13 June 2014.

5. What is a serious challenge for national security and defense planners is an opportunity for analysts!

6. Andrew L. Ross, "The Dynamics of Military Technology," in David Dewitt, David Haglund, and John Kirton, eds., *Building a New Global Order: Emerging Trends in International Security*, Toronto: Oxford University Press, 1993, pp. 106–140.

7. Nanotechnology appears to be revolutionizing materials science and engineering.

8. Think suicide bombing and IEDs. On the former see Michael C. Horowitz, "Nonstate Actors and the Diffusion of Innovations: The Case of Suicide Terrorism," *International Organization*, Vol. 64, No. 1, Winter 2010, pp. 33–64.

9. The Preparatory Commission for the CTBTO, for instance, which has developed an International Monitoring System (IMS) and Global Communications Infrastructure (IGC) to support implementation of the CTBT. See also the NATO Multiple Futures Project @ http://www.act.nato.int/mainpages/multiplefutures, accessed 26 December 2012.

10. Emily O. Goldman and Andrew L. Ross, "Conclusion: The Diffusion of Military Technology and Ideas—Theory and Practice," in Emily O. Goldman and Leslie C. Eliason, eds., *The Diffusion of Military Technology and Ideas*, Stanford: Stanford University Press, 2003, pp. 371–403; and Michael C. Horowitz, *The Diffusion of Military Power: Causes and Consequences for International Politics*, Princeton: Princeton University Press, 2010.

11. If unevenly.

12. The advantage enjoyed by technological leaders may not necessarily be fleeting, but it cannot be expected to be permanent.

13. What R&D programs to invest in, what to produce and procure, now much to procure.

14. For insightful recent surveys see Shawn Brimley, Ben FitzGerald, and Kelley Sayler, *Game Changers: Disruptive Technology and U.S. Defense Strategy*, Washington, DC: Center for a New American Security, September 2013; and Michael E. Horowitz, "Coming Next in Military Tech," *Bulletin of the Atomic Scientists*, Vol. 70, No. 1, January 2014, pp. 54–62. See also James D. Shields and James A. Tegnelia, Co-Chairmen, Defense Science Board Report on *Technology and Innovation Enablers for Superiority in 2030*, Washington, DC: Office of the Under Secretary of Defense for Acquisition, Technology, and Logistics, October 2013.

15. See Committee for Science and Technology Challenges to U.S. National Security Interests, Division on Engineering and Physical Sciences, National Research Council of the National Academies, *Report of a Workshop on Big Data*, Washington, DC: The National Academies Press, 2012; Chris A. Mattmann, "A Vision for Data Science," *Nature*, Vol. 493, 24 January 2013, pp. 473–475.

16. Clay Wilson, "Avatars, Virtual Reality Technology, and the U.S. Military: Emerging Policy Issues," CRS Report for Congress RS22857, Washington, DC: Congressional Research Service, 9 April 2008.

17. James Manyika, Michael Chui, Jacques Bughin, Richard Dobbs, Peter Bisson, and Alex Marrs, *Disruptive Technologies: Advances That Will Transform Life, Business, and the Global Economy*, McKinsey Global Institute, McKinsey & Company, May 2013, pp. 69–78; Robert O. Work and Shawn Brimley, *20YY: Preparing for War in the Robotic Age*, Washington, DC: Center for a New American Security, January 2014.

18. Kevin Y. Ma, Pakpong Chirarattananon, Sawyer B. Fuller, and Robert J. Wood, "Controlled Flight of a Biologically Inspired, Insect-Scale Robot," *Science*, Vol. 340, No. 6132, 3 May 2013, pp. 603–607; Ron Cowen, "Tiny Robot Flies Like a Fly," nature.com, 2 May 2013, available @ http://www.nature.com/news/tiny-robot-flies-like-a-fly-1.12926, accessed 14 May 2013; "Miniature Flying Robots: Robodiptera," *The Economist*, 4 May 2013, p. 77; Elisabeth Bumiller and Thom Shanker, "War Evolves With Drones, Some Tiny as Bugs," *The New York Times*, 20 June 2011.

19. Manyika et al., *Disruptive Technologies*, pp. 79–86.

20. Precision strike is an extensively employed, much-exercised capability for some; a new, emerging capability for others. Randy Huiss, *Proliferation of Precision Strike: Issues for Congress*, CRS Report for Congress R42539, Washington, DC: Congressional Research Service, 14 May 2012; Thomas G. Mahnken, "Weapons: The Growth & Spread of the Precision-Strike Regime," *Daedalus*, Vol. 140, No. 3, Summer 2011, pp. 45–57.

21. "Energy Weapons: Zap, Crackle and Pop," *The Economist*, 1 September 2012, pp. 12–23.

22. A counter to unmanned systems and precision-strike weapons?

23. James M. Acton, *Silver Bullet? Asking the Right Questions about Conventional Prompt Global Strike*, Washington, DC: Carnegie Endowment for International Peace, 2013.

24. A capability not yet exhibited in warfare by states in Asia and the Pacific.

25. BMD is perennially new and emerging.

26. Drawn from National Intelligence Council, *Global Trends 2030: Alternative Worlds*, pp. 99–100. See "Special Section: Biomaterials," *Science*, Vol. 338, No. 6109, 16 November 2012, pp. 889–926; and Yoskhiki Sasai, "Grow Your Own Eye," *Scientific American*, Vol. 307, No. 5, November 2012, pp. 44–49.

27. David Ewing Duncan, "How Science Can Build a Better You," *The New York Times*, 4 November 2012; Committee on Assessing Foreign Technology Development in Human Performance Modification, Division on Engineering and Physical Sciences, Board on Behavioral, Cognitive, and Sensory Sciences, Division on Behavioral and Social Sciences and Education, *Human Performance Modification: Review of Worldwide Research with a View to the Future*, Washington, DC: The National Academies Press, 2012.

28. Michael Behar, "Invasion of the Body Hackers," *The New York Times Magazine*, 25 May 2014, pp. 36–41 & 59.

29. David Pogue, "The Remote Control in Your Mind," *Scientific American*, Vol. 307, No. 6, December 2012, p. 32; Nick Bilton, "Computer-Brain Interfaces Making Big Leaps," *The New York Times*, 4 August 2013, available @ http://bits.blogs.nytimes.com/2013/08/04/disruptions-rather-than-time-computers-might-become-panacea-tohurt/, accessed 2 April 2014; Peter J. Ifft, Solaiman Shokur, Zheng Li, Mikhail A. Lebedev, and Miguel A. L.

Nicolelis, "A Brain-Machine Interface Enables Bimanual Arm Movements in Monkeys," *Science Translational Medicine*, Vol. 5, No. 210, 6 November 2013, pp. 1–3; Kim Tingley, "The Body Electric," *The New Yorker*, 25 November 2013, pp. 78–86.

30. To "provide enhanced experiences of real-world situations." National Intelligence Council, *Global Trends 2030*, p. 100.

31. "A Third Industrial Revolution," *The Economist*, 21 April 2012, available @ http://www.economist.com/node/21552901?fsrc=nlw|wwp|12-27-2012|4467601|36961439|; accessed 27 December 2012; Matthew Hallex, "Digital Manufacturing and Missile Defense," *Public Interest Report*, Vol. 66, No. 2, Spring 2013, available @ http://blogs-cdn.fas.org/pir/wp-content/uploads/sites/8/2013/05/Digital-Manufacturing-and-Missile-Proliferation-Spring-13.pdf, accessed 28 May 2013. Manyika et al., *Disruptive Technologies*, pp. 106–114. See also National Intelligence Council, *Global Trends 2030*, p. 93.

32. Lauren Biron, "Navy's Future: Unmanned Vehicles, Robots and 3D Data," C4ISR&NETWORKS (c4isrnet.com), 3 July 2014, available @ http://www.c4isrnet.com/article/20140702/C4ISRNET08/307020007/Navy-s-future-Unmanned-vehicles-robots-3D-data, accessed 4 July 2014, provides a snapshot of emerging US Navy technologies.

33. For an accessible attempt, by a former editor of *Nature*, to anticipate what discoveries might lie ahead, see John Maddox, *What Remains to Be Discovered: Mapping the Secrets of the Universe, The Origins of Life, and the Future of the Human Race*, New York: Touchstone, 1999. See also the special feature on "The Future in 50, 100 and 150 Years," *Scientific American*, Vol. 308, No. 1, January 2013; and "World Changing Ideas," *Scientific American*, Vol. 307, No. 6, December 2012, pp. 34–45.

34. Jonathan Adams, "The Rise of Research Networks," *Nature*, Vol. 490, 18 October 2012, pp. 335–336; Jonathan Adams, "The Fourth Age of Research," *Nature*, Vol. 497, 30 May 2013, pp. 557–560.

35. The future of scientific inquiry belongs to those who frame it.

36. Manyika *et al.*, *Disruptive Technologies*, pp. 87–95; Kenneth Ford and Clark Glymour, "The Enhanced Warfighter," Bulletin of the Atomic Scientists, Vol. 70, No. 1, January 2014, pp. 43–53.

37. Andrew Pollack, "Scientists Add Letters to DNA's Alphabet, Raising Hope and Fear," The New York Times, 7 May 2014, available @ http://www.nytimes.com/2014/05/08/business/researchers-report-breakthrough-in-creatingartificial-genetic-code.html, accessed 8 May 2014; Ewen Callaway and Nature, "First Life with 'Alien' DNA Created in Lab," scientificamerican.com, 7 May 2014, available @ http://www.scientificamerican.com/article/first-life-withalien-dna-created-in-lab/, accessed 8 May 2014; "DIY Chromosomes," *The Economist*, 29 March 2014, p. 83; Laurie Garrett, "Biology's Brave New World: The Promise and Perils of the Synbio Revolution," Foreign Affairs, Vol. 92, No. 6, November/December 2013, pp. 28–46; Richard Lewontin, The New Synthetic Biology: Who Gains?" *The New York Review of Books*, 8 May 2014, pp. 22–23.

38. Rodney Brooks, "The Merger of Flesh and Machine," in John Brockman, ed., *The Next Fifty Years: Science in the First Half of the Twenty-First Century*, New York: Vintage Books, 2002, pp. 183–193.; Gretchen Vogel, "Human

Stem Cells Form Cloning, Finally," *Science*, Vol. 340, No. 6134, 17 May 2013, p. 795.

39. James Gorman, "Brain-Mapping Milestones," *The New York Times*, 21 April 2014, available @ http://www.nytimes.com/2014/04/22/science/brain-mapping-milestones.html, accessed 23 April 2014; James Gorman, "Brain Control in a Flash of Light," *The New York Times*, 21 April 2014, available @ http://www.nytimes.com/2014/04/22/science/mind-control-in-a-flash-of-light.html, accessed 23 April 2014.

40. John Markoff, "Scientists See Promise in Deep-Learning Programs," *The New York Times*, 23 November 2012.

41. Quentin Hardy, "A Big Leap to Quantum Computing," *The New York Times*, 13 May 2013; Quentin Hardy, "Testing a New Class of Speedy Computer, *The New York Times*, 22 March 2013; Adrian Cho, "New Form of Quantum Computation Promises Showdown With Ordinary Computers, *Science Now*, 21 December 2012, available @ http://news.sciencemag.org/sciencenow/2012/12/new-form-of-quantum-computation-.html?rss= 1, accessed 26 December 2012; Brad Hooker, "In Depth: Quantum Computing: The Next Information Revolution," *AAAS Member Central*, 16 November 2012, available @ http://membercentral.aaas.org/blogs/depth/quantum-computing-next-information-revolution, accessed 30 November 2012; Jane Qiu, "Quantum Communications Leap Out of the Lab," *Nature*, Vol. 508, 24 April 2014, pp. 441–442; John Markoff, "Microsoft Makes Bet Quantum Computing Will Be Next Big Leap," *The New York Times*, 23 June 2014, available @ http://www.nytimes.com/2014/06/24/technology/microsoft-makes-a-bet-on-quantum-computing-research.html, accessed 24 June 2014.

42. Manyika et al., *Disruptive Technologies*, pp. 115–125; Nir Shitrit et al., "Spin-Optical Metamaterial Route to Spin-Controlled Photonics," *Science*, Vol. 340, No. 6133, 10 May 2013, pp. 724–726; special feature on "The Future of Manufacturing," *Scientific American*, May 2013, pp. 36–51; Steven Ashley and Larry Greenemeier, "9 Materials That Will Change the Future of Manufacturing," scientificamerican.com, 22 April 2013, available @ http://www.scientificamerican.com/article.cfm?id= 9-materials-that-will-change-manufacturing, accessed 15 May 2013.

43. Clive Hamilton, "Geoengineering: Our Last Hope, or a False Promise?" *The New York Times*, 26 May 2013, available @ http://www.nytimes.com/2013/05/27/opinion/geoengineering-our-last-hope-or-a-false-promise.html?pagewanted= all, accessed 27 May 2013; Clive Hamilton, *Earthmasters: The Dawn of the Age of Climate Engineering*, New Haven: Yale University Press, 2013; Clive Hamilton, "Geoengineering and the Politics of Science, *Bulletin of the Atomic Scientists*, Vol. 70, No. 3, May 2014, pp. 17–26; "Scientist Suggests Refreezing the Arctic to Stop Global Warming," *The Economic Times*, 12 December 2012, available @ http://economictimes.indiatimes.com/environment/global-warming/scientist-suggests-refreezing-the-arctic-to-stop-global-warming/articleshow/17587125.cms, accessed 28 December 2012.

44. On computer science and complexity see John Pavlus, "Machines of the Infinite," *Scientific American*, Vol. 307, No. 3, September 2012, pp. 66–71.

45. W. Pfaff et al., "Unconditional Quantum Teleportation between Distant Solid-State Quantum Bits," Sciencexpress, 29 May 2014, pp. 1–8, available

@ https://www.sciencemag.org/content/early/2014/05/28/science.1253512. full.pdf, accessed 5 July 2014.

46. Moti Fridman, Alessandro Farsi, Yoshitomo Okawachi, Alexander L. Gaeta, "Demonstration of Temporal Cloaking," *Nature*, Vol. 481, 5 January 2012, pp. 62–65.

47. This material is drawn and adapted from Andrew L. Ross, "Framing Chinese Military Innovation," in Tai Ming Cheung, ed., *China's Emergence as a Defense Technological Power*, London and New York: Routledge, 2013, pp. 187–213.

48. On technology and the technological component of innovation, see Colin S. Gray, "Technology as a Dynamic of Defence Transformation," *Defence Studies*, Vol. 6, No. 1, March 2006, pp. 26–51; and Ross, "The Dynamics of Military Technology."

49. The distinction between "hardware" and "software" employed here is drawn from Ross, "The Dynamics of Military Technology," pp. 106–140.

50. China's People's Liberation Army, for instance, is in the process of learning how to use its much ballyhooed aircraft carrier—a fine example of cutting-edge 1980s technology, as Richard Bitzinger put it—and other recently acquired technological/hardware capabilities. Kathrin Hille, "Military Machine: Learning to Use New Equipment," *Financial Times*, 11 December 2012.

51. For an example of work on organizational innovation see Chad C. Serena, *A Revolution in Military Adaptation: The US Army in the Iraq War*, Washington, DC: Georgetown University Press, 2011.

52. For examples of work on operational innovation see Lazar Berman, "Capturing Contemporary Innovation: Studying IDF Innovation against Hamas and Hizballah," *Journal of Strategic Studies*, Vol. 35, No. 1, February 2012, pp. 121–147; Meir Finkel, *On Flexibility: Recovery from Technological and Doctrinal Surprise on the Battlefield*, Stanford: Stanford University Press, 2011; John A. Nagl, *Learning to Eat Soup with a Knife: Counterinsurgency Lessons from Malaya and Vietnam*, Chicago: University of Chicago Press, 2002; James A. Russell, *Innovation, Transformation, and War: Counterinsurgency Operations in Anbar and Ninewa Provinces, Iraq, 2005–2007*, Stanford: Stanford University Press, 2011; David H. Ucko, *The New Counterinsurgency Era: Transforming the U.S. Military for Modern Wars*, Washington, DC: Georgetown University Press, 2009.

53. As Fred Iklé aptly observed, "Military services cling to the types of weapons to which they have become accustomed, seeking marginal improvements rather than radical innovation." Fred Charles Iklé, "Can Nuclear Deterrence Last Out the Century?" *Foreign Affairs*, Vol. 51, No. 2, January 1973, p. 384. For an insightful examination of the relationship between organizations, or institutions, and technology, see Timothy Moy, *War Machines: Transforming Technologies in the U.S. Military, 1920–1940*, College Station: Texas A&M University Press, 2001.

54. The distinction between sustaining and disruptive innovation employed here is drawn from Clayton M. Christensen, *The Innovator's Dilemma*, New York: HarperBusiness, 2000. Peter Dombrowski and Eugene Gholz, "Identifying Disruptive Innovation: Innovation Theory and the Defense Industry," *Innovations*, Vol. 4, No. 2, Spring 2009, pp. 101–117, further develop the concept of disruptive innovation.

55. A point also emphasized by Dima Adamsky, *The Culture of Military Innovation: The Impact of Cultural Factors on the Revolution in Military Affairs in Russia, the US, and Israel*, Stanford: Stanford University Press, 2010, p. 1.

56. See, for instance, the roster of military revolutions provided by Andrew F. Krepinevich, "Cavalry to Computer: The Pattern of Military Revolutions," *The National Interest*, No. 37, Fall 1994, pp. 30–42.

57. Krepinevich, "Cavalry to Computer: The Pattern of Military Revolutions;" and MacGregor Knox and Williamson Murray, eds., *The Dynamics of Military Revolution, 1300–2050*, Cambridge: Cambridge University Press, 2001.

58. Mark Stefik and Barbara Stefik, *Breakthrough: Stories and Strategies of Radical Innovation*, Cambridge: The MIT Press, 2004, p. 3.

59. As R. M. Henderson and K. B. Clark, "Architectural Innovation: The Reconfiguration of Existing Product Technologies and the Failure of Established Firms," *Administrative Science Quarterly*, Vol. 35, No. 1, 1990, pp. 9–30, put it, "The essence of an architectural innovation is the reconfiguration of an established system to link together existing components in a new way" (p. 12).

60. Simultaneous hardware and architectural breakthroughs actually appear to be the exception rather than the rule. Breakthrough doctrinal and organizational innovations often lag breakthrough technological innovations.

61. On the significance of performance metrics see Peter Dombrowski, Eugene Gholz, and Andrew L. Ross, *Military Transformation and the Defense Industry after Next: The Defense Industrial Implications of Network-Centric Warfare*, Newport Paper No. 18, Newport, RI: U.S. Naval War College Press, 2003; and Peter Dombrowski and Eugene Gholz, *Buying Military Transformation: Technological Innovation and the Defense Industry*, New York: Columbia University Press, 2006.

3
Absorbing New Military Capabilities: Defense Technology Acquisition and the Asia-Pacific

Martin Lundmark

Sweden has been an oft-cited example of a smaller nation that has produced a wide assortment of cutting-edge defense systems—many times producing substantial advances in emerging technologies. The Swedish defense companies rank among the most R&D-intensive in Sweden's overall very export-oriented industry. Saab, which accounts for more than 80% (2014) of the aggregate defense turnover in Sweden, had an R&D share of 25 percent (2014). Of this R&D, 30 percent is internally generated funds. The Swedish defense export has roughly tripled in the last ten years, Sweden is the twelfth biggest defense exporter globally in the period 2003–2012 and it is the nation with the highest defense export per capita.[1] The Swedish defense companies on average have a 60 percent export share, and around 20 percent R&D as share of turnover.[2] This points to the fact that defense products developed in Sweden receive considerable demand globally—these products are however increasingly developed in international innovation networks. Sweden is just an example of a nation with a developed defense innovation system that Asia-Pacific nations could collaborate with.

Asia-Pacific nations in general have a much shorter history of domestic defense innovation and a less developed domestic defense innovation infrastructure. They could learn from the Swedish example how to organize and implement an efficient defense innovation process, and they could also find mutually beneficial (for the military, government, industry and academia) interaction through partnering with defense-exporting nations.

This chapter will attempt to exemplify the ongoing transformation of Swedish defense acquisition and its relation to the domestic defense industry. Sweden gradually in the years 1997–2013 decreased its autarky ambitions, going in the opposite direction compared to certain Asia-Pacific nations like South Korea, China, and India. However, since 2013, Sweden has declared two "vital strategic interests": fighter aircraft capability (2013) and underwater capability (2014). The Swedish example is used in order to suggest how Asia-Pacific nations can partner with certain defense-technology-savvy developed nations. In this chapter, "emerging technologies" are treated as military capabilities in the Asia-Pacific region that are new to a nation, and possibly to the region. The purpose of this chapter is to suggest a systematic approach for creating a new military capability in a nation through import of defense technology and the bilateral partnering that comes with it. In doing this, the chapter aims to respond to the following questions:

- What are the sources of input to national defense innovation systems?
- How can a nation analyze how to create a new military capability through defense materiel import?
- How can a *technology policy* and *bilateral partnering* help to improve the process of developing new military capabilities in a nation?

Emerging technologies

Few, if any, nations have independent, indigenous defense technology development; they are in parts or mostly dependent on inflow of technologies. Nations also must be cost-efficient and assure technology access for a long period of time. If they want to truly access "emerging" technologies, this brings with it considerable economic and technology risk. If they cannot access emerging technologies through indigenous development, they must find other ways. They may acquire existing, operative defense materiel.

Companies can also strive to become parts of internationalized networks and collaboration networks. Such networks could be through bi-, tri-, or multilateral collaborative projects, for example, through a joint development of a missile, ship, radar, or other weapon systems. Concept pairs as continuous-discontinuous, sustaining-disruptive, incremental-transformational, minor-radical, and evolutionary-revolutionary constitute tools for describing the "continuum of military innovation." Most military innovation is continuous and undramatic, and can be handled through established routines.[3] The concept "emerging technologies" denotes that the technology is immature, new, exploratory, and with an uncertain future and application. Thereby the challenges and uncertainties become larger and harder to plan for when developing a new military capability.

The US is one of the very few nations—perhaps the only one—that can invest broadly in most technologies, and finance competing corporate structures as in the case of the competition between Lockheed Martin and Boeing for Joint Strike fighter. For a technology leader like the US, the technological risks taken become immense. For an Asia-Pacific nation acquiring new technology in order to radically increase its military capability—or add a new one—the challenges to reach that capability become substantial. It must possess an ability to exploit and operate the possibilities of this new military capability.

National defense innovation

Innovation within a nation can be said to exist in a combination of technologies, institutions, and organizations: what can be labelled "systems of innovation." Through this view, innovation and technological change can be studied as a source of continuous development. Innovations are for the most part new combinations of existing elements. This process is characterized by complicated feedback mechanisms and interactive relations involving science, technology, learning, production, policy, and demand.[4] The "triple-helix" perspective points to the synergistic effect of the relations of university–industry–government, and enhances the importance of the university sector as having a central role in national innovation.[5] An issue relevant for the transfer of technology and the build-up of military capabilities is the issue of "absorptive capacity;" to what extent a company or an organization has the capacity to make use of and exploit technologies and its operative use.[6] For the analysis of military innovation and how it is intended to develop a nation's military capabilities, a military innovation triad perspective can be used with three components that create the conditions for military innovation: technology, doctrine, and organization.

For the nation aspiring to attain a new military capability, a consideration must be made regarding the technology-absorption capacity of indigenous companies, concerned organizations (e.g. in procurement and research), and most of all the military. If the domestic defense technology infrastructure is not sufficiently sophisticated and adaptive, there might be a tremendous challenge to implement the intended capability addition. Ideally, a synergistic effect should be attained through the triple-helix effect of the combined contributions of academia, industry, and government. In the previous paragraph there were different emphases in the analysis of innovation: the systems of innovation model seeing the state as having the leading role; the triple-helix model having academia as the central enabler of innovation. The most obvious perspective would be to see defense companies as the nexus of innovation.

Another perspective on the sources of innovation is the increasing importance of non-defense technology development for military innovation—this will be discussed below under the heading "globalization in networks."

In the perspective of this chapter, "government" consists of the military on one side and other defense-related government agencies on the other (e.g., defense research, procurement, testing facilities). Since the role distribution and separation between the military and various government agencies highly differ between nations, an analysis of specific nations' defense innovation infrastructure would be needed in order to understand its dynamics. An importing nation can together with the exporting nation interact and collaborate on the development of the new military capability. In each nation there will be a nation-specific composition of: defense industry or industry that can participate in manufacturing or development of defense materiel; a military organization; certain defense-related government agencies; academic institutions; and R&D institutes. Together with these two separate infrastructures, each nation will have a certain military doctrine.

The two nations are connected by the technology and the defense systems that are to become a part of the importing nation's doctrine, form the basis of a new capability, and be used by the military. A fundamental factor for such a bilateral export partnership becoming constructive and successful is the perceived fit between the two national infrastructures and the military doctrine that guides them. If the defense companies in the respective nations see a fit between the project goals and their corporate long-term strategic goals, the probability for success is much higher. Reversely, if a collaborative structure is forced upon the companies by government actors (either military or defense ministries), the collaboration runs a much higher risk of being cancelled or running into implementation problems.[7]

Technology policy

A nation or an organization must define its relationship to the constant flow of technology, to innovation, and to how its ambitions and resource conditions must harmonize with its actions. The following discussion concerning a technology policy for military technology acquisition is partly based upon studies in 2011 and 2012 for the Swedish Ministry of Defense (MoD) regarding a technology policy for C3I in the land arena.[8]

In order to transform and reposition a national defense technology acquisition, one tool is to define a *technology policy* that reflects the position a nation (or a company) wants to achieve and perfect. A *policy* constitutes fundamental principles for action and is therefore a framework for decision-making. A technology policy can be described as principles for choices of technology and related questions such as levels of knowledge, levels of investment, frequency of renewal, implementation processes, and organizational demands. A technology policy thus comprises a portfolio of choices—of decisions—that will enable the fulfillment of goals and also deal with upcoming threats and opportunities.[9]

Technology refers to the knowledge of techniques. Technology denotes the art of mastering techniques. Technologies relate to each other in different ways.[10] Technology exists and acts in an ever-evolving interaction with military doctrine. Doctrine may push development of further technology perfection that enables certain capabilities, but technology may also offer new doctrinal possibilities.[11]

The starting point for a technology policy should be an overarching direction and relationship for what position that is desired vis-à-vis the relevant parts of the environment—given the limitations of the accessible resources (financial, organizational, and competence-wise). It is thus not meaningful to formulate a policy that is not realistic. For a national government, or its armed forces, the technology policy strives to take a comprehensive view in order to reach positive effects that otherwise might be overlooked. It is also fundamental to strive for a cost-effective technology acquisition. In the formulation of a technology policy, the following six areas are central:[12]

1. *Technology areas, specializations, and their integration into the systems where they should bring effects*: In order to define the level of ambition, a suitable taxonomy is to use five "performance objectives:" *quality, delivery assurance, speed, flexibility*, and *cost*. These five must form a balance, and be in sync with what the military needs and what funds that are at hand.[13]
2. *Proximity to the development front*: for example, ambition level for the technology developments. How to combine and organize efforts at different technology-development levels.

3. *Sources of technological capability*: to what extent do you possess the sufficient and relevant technology competence? Sources could be, for example, domestic, imported, academia, collaborative partners, network partners.
4. *Level of investment in R&D*: for example, technology areas, investment in organization and competence, facilitating resources, degrees of flexibility, and risk-taking.
5. *Timing of investment*: When should you invest in a technology; for example, be technology leader, early adopter, follower, or late follower?
6. *Organization*: What should the domestic institutional structures be, and what should concerned organization's roles be? How do you structure processes in and between institutions? What are the incentive structures? How should certain decisions and choices be made? How will intellectual property be handled, shared, and protected?

These six areas are not independent; if the conditions are altered in any of them they will affect the other areas. How they relate varies depending on ambition, resources, and the demands given by the environment. The six areas must form a balanced wholeness; and decisions in one area (in the above order) set conditions for the next. The military procurement, capability creation, and sustainment must over time have a balance vis-à-vis its conditions, and the technology policy offers a tool for creating such a balance. Some of the most frequent problems are cost over-runs, delayed delivery, and that the delivered product does not correspond to the defined quality specifications. A technology policy delimits the risk for such downfalls.

Swedish transformations

In order to set the use of a technology policy in perspective, Sweden's changes in defense matters in the last decades will be discussed. Three aspects will be focused upon:

- The repositioning of the Swedish defense posture
- The present transformation of Swedish defense acquisition
- The development of Swedish defense companies' exports

Repositioning of Sweden's defense posture in the last decades

Sweden, as many other nations, has had several upheavals and transformations in its military structure, size, doctrine, and not the least its relationship to other nations and actors in other nations. Sweden had during the Cold War a high level of self-reliance in defense materiel and an unusually large and sophisticated defense industry compared to the size of the nation. The threat assessment was stable with the Soviet Union just over the Baltic, and Sweden had, despite its non-membership of NATO, favorable defense technology relations with the US. Sweden had ambitious standards for its domestic defense-technology development, and has been repeatedly studied as an interesting example of a smaller nation with a surprisingly advanced defense industry.

In lacking a clear threat after the Cold War, a search for a new threat assessment occupied many minds. Around the turn of the century, Sweden whole-heartedly embraced network-based defense and reached global cutting-edge prominence on network-enabled capabilities (NEC)—at least on power-point slides. These ambitions, however, came to very little compared to the very large funds that for a few years were directed towards NEC.

During the first ten years of the new millennium Sweden dramatically redirected its military focus towards international peacekeeping operations, where Sweden has participated in Afghanistan under NATO coordination; in and around Africa under UN coordination (e.g. Ethiopia, Libya, Somalia, Sudan, and Chad) and also in the Eastern Mediterranean. The homeland defense that had been the focus during the Cold War became less prioritized for a few years. In recent years, the pendulum has swung back, and presently there is more focus on the capabilities for defending the Swedish territory. The recent ambitious modernization of the Russian Armed Forces and defense industry[14] has further emphasized the focus on homeland defense, and can be understood as a strong incentive for the decision in August 2012 to upgrade the Gripen fighter to the E/F version[15] and to develop a new, indigenous submarine (the Type A26).[16]

From the mid-1990s onwards, Sweden has also increasingly engaged in multilateral arms collaboration. Sweden had very limited experience of this, compared to the dominating European collaborating nations (that had gradually and increasingly developed their collaborative efforts since the 1950s).[17] Naturally, this transition did not come easily. During the same period, many Swedish defense companies were also acquired abroad—Sweden was very liberal compared to other nations in allowing foreign acquisition of defense companies.

The previous long-lasting tradition of sophisticated domestic arms development, followed by détente, multilateral arms collaboration, internationalized defense industry, international operations, and then back to more focus on homeland defense has in Sweden brought with it strong friction between indigenous traditions and the constant but changing demand for redirection of military policy priorities. The Swedish defense innovation infrastructure and doctrine has thus experienced considerable shifts of balances.

Present transformation of Sweden's defense acquisition

In recent years the MoD has chosen not to establish a defense-industrial policy, something that most European states have established. A defense acquisition strategy has however been established (from the Armed Forces in 2007, and the MoD in 2009). This strategy declares that Sweden should as a *first* priority aim to upgrade existing defense materiel, *secondly* to acquire already operative materiel ("off-the shelf"), *thirdly* to develop in collaboration with other nations, and *fourthly* (and exceptionally) develop indigenously without foreign partners. This constitutes a paradigm shift with previous traditions—a shift that is further supported by current EU reforms towards creating an open and level EU defense marketplace.

The Swedish MoD very clearly stated in a 2009 government bill that in the C3I Land Arena (the army and resembling units in the air force and navy) the most important focus must be that the defense materiel must be operative: operative *on time* when certain capabilities and units are promised to be operative; and also that technology ambitions should stay at "*good enough.*" In order to fulfill this, the C3I Land Arena must acquire solutions that are already operative in other nations, and *not* develop indigenous solutions. This was formulated by the MoD in a very straightforward manner, and reflected the uncertainties that prevailed in this area. The most striking change that this imposes upon the armed forces and the procurement agency concerns the issue of timing for investment; that the position must change from a self-image of being a technology leader (without the funds to go with it, however) into becoming a *late follower* of the technology development.

Dominant technologies should be sought. This shift of position primarily means that Sweden should acquire, for example, radio, communication, and phone solutions that are already in use in similar nations. A clear problem here is that there are several immature technologies (e.g. broadband width and digital radio: Joint Tactical Radio System (JTRS)) and there are in several areas no clear, dominating technologies or companies. But with no funds for domestic new development, this must be sought. What we discovered in our study was that there were a number of acquisition programs that still were not delivered that had been initiated during periods under a different military policy and different conditions for domestic defense industry. With a distinct shift towards expeditionary forces and operative products and capabilities *on time*, this constitutes a big challenge.

Sweden presently strives to find economies of scale and scope through adopting existing standards and trying to create shared, multilateral acquisitions or the latest fad: pooling and sharing. In parallel, Sweden and the exporting companies try to find win-win technology synergies with export customers.

Every nation must have a technology and product inflow of defense technology. Not even the US is self-reliant in defense technology. The US is also dependent upon access to certain critical components (e.g. components which performance is optimized towards what can be achieved through the use of rare earth elements) that are not produced in the US, and that are typically produced in China.

Sweden's foremost priority as defined by the MoD in 2009 (and repeatedly iterated since) is thus that the priority of Swedish defense procurement is to predominantly procure mature and proven technologies and defense materiel. An importing nation by definition imports an already developed defense technology. The technology and the defense system must in all import cases be modified in order to fit with the importing nation's specific demands, doctrine, and existing defense infrastructure. In this modification process, the separate domestic defense innovation infrastructures must interact.

Development of the Swedish defense export

As stated previously, the defense companies for a long time predominantly produced and sold to the Swedish Armed Forces. In the last 15 years, three important changes have been put upon the defense industry in Sweden:

1. Several Swedish defense companies have been acquired by foreign companies (Hägglunds, Kockums, and Bofors) in the period 1996–2005.
2. Sweden to a much lesser extent indigenously develops its defense materiel, and also increasingly acquires defense materiel off-the-shelf.
3. Defense R&D has decreased by more than 50 percent in the last 7 years (due to shifts of funds to international operations).

These three strands of development could intuitively point to the fact that the Swedish defense companies should have faced a gradual decrease of competitiveness and attractiveness. However, the overall defense export increased over the period 2002–2011. Sweden has become the number one defense exporter per capita in the world; Sweden was the world's tenth largest defense exporter in the period 2002–2011; the defense export tripled in the period 2002–2011.[18] This export-based position may well be questioned and criticized, but what it nevertheless clearly suggests is that the Swedish defense companies' products meet considerable demand in the global marketplace.

Sweden has chosen to largely abandon indigenous development in certain technology areas (especially radar, battle tanks, and missiles) whilst companies that sell such products receive export support. In several other technology areas the ambitions are lowered and/or are supported through multilateral arms collaboration (e.g. Meteor, Iris-T, Neuron). In only two areas is there still a firm commitment to cutting edge development: submarines and fighters. Regarding submarines, there is limited collaboration, and Kockums was for a long time to some extent strategically hamstrung through being owned by the main European competitor ThyssenKrupp Marine Systems. However, Kockums was acquired by the Swedish defence company Saab in 2014. Added to the submarines, torpedo development is also performed domestically, tailored to Swedish military demand. Regarding fighters (i.e., Saab 39 Gripen), there was a major decision in August 2012 to upgrade 40 to 60 Gripen version C/D to E/F. This upgrade was made possible through Switzerland buying 22 Gripen; the MoD had in 2011 set a condition that an upgrade to E/F could only occur if "Brazil, or another nation acquires Gripen."[19] It appears, however, that despite the lowered ambitions in defense materiel development, Sweden's national defense innovation system has withheld its aggregate ability for system integration.

For the defense companies in Sweden, export has become much more important. Most of them had very little export until the early 1990s. By now the average export share has grown to around 70 percent, in certain areas close to 100 percent.[20] In most cases, the defense export is associated with a defense offset obligation. Such an offset obligation typically concerns 150–200 percent of the order value and also, increasingly, complex structures of technology transfer to defense companies in the buying nations. A company that cannot convince prospective customers that its offset and technology transfer design is competitive in most cases does not stand a chance against its competitors. Defense export typically brings with it long-term bilateral collaborative structures (often ten years) across the entire spectrum of military, government, research and corporate organizations.[21] Thus, defense export has, compared to domestic technology development, become much more important for the companies, in Sweden and in several other nations. The large increase in export and export share and the foreign ownership paired with the increased defense materiel development collaboration has in aggregate made defense companies in Sweden globalized into international networks.

Globalized production in networks

A development of a new military capability concerns many years of development and use. Issues of life cycle costs have come more in focus, at the same time as being difficult to calculate in a satisfactory way. As the life of a product or an installation extends, it becomes increasingly elusive to predict how it will be used or how the product itself will behave. Innovation in general, and also in defense, is becoming increasingly dispersed into cross-border networks. Companies must create creative links with innovative and cooperative partners that can maximize revenues at the same time as controlling risk and uncertainty; in order to manage complex cooperation and innovation, "strategic sourcing" has become more important.[22]

The conditions for production of defense materiel have fundamentally been altered through the general globalization of production in society. Highly sensitive defense systems like precision-guided munitions, sensors, and electronic warfare depend upon the access to components that are produced in China—and it is not economically feasible, for example, to produce certain components based upon rare earth metals outside China; it would not be at all profitable. However, China is dependent upon an undisturbed flow of goods to industrialized nations, and most such components are not defense-specific. For European and US defense companies the design and system integration into the final defense system remains with them. China, however, steadily moves upwards in the complexity hierarchy and strives to be able to produce defense equipment of ever-higher sophistication.[23]

The technological building stones of the "RMA, NEC, Transformation," and so on, are more elusive than the technologies that underpinned the defense innovation a few decades ago. A well-known fact is that defense innovation no longer leads technology development in general, and that all production has become more globalized. Furthermore, the cycle times and life expectancy of each technology generation is becoming shorter and shorter.

However, defence systems are first, planned for several years, procured for some more years, delivered and tested for some period, and finally in service (with upgrades) for a decade or two. With an emerging technology, there are inherent uncertainties and hence profound technological and economic risks. There is thus a clear mismatch between the overarching planning and implementation cycles of defense procurement compared to generic technology development. For example, the Swedish Armed Forces had communication capability uncertainties a few years ago and invested strongly in funds and planning into the US-led JTRS, and when this program was abruptly cancelled in October, 2011, a new start had to be taken. If compared to a technology policy, Sweden suffered the consequences of investing into immature technologies. A problem in state-of-the-art C3I is that there is no dominant technology, and no mature solutions. So what should really be followed?

Defense procurement must become more flexible and have shorter cycles, since non-defense production and innovation defines the pace and will not accommodate to defense incentives. A nation like the US will continue to be the foremost innovator in defense, but lesser nations must accept that they should be late adopters overall, and early followers at best. In narrow niches (e.g., cryptography, electronic warfare, countermeasures, C3I system integration) they might be forced to find nation-specific solutions due to arms legacy and national organization solutions. For a cost-efficient defense procurement process, the guiding principles for technology priorities should in most nations be "good enough," proven technologies, and assurances of operative in time. For an Asia-Pacific nation aspiring for new military capabilities, the good-enough posture, or at least not-the-technology-leader posture, should be taken. By importing and modifying existing defense technology, this becomes the case. The often high-strung rhetoric of the capability to be acquired must, however, reach its pragmatic solution on the ground.

Implications and relevance to the Asia-Pacific region

So what about the importance of a technology policy, emerging technologies, and the military situation in the Asia-Pacific region? If we assume that a certain number of states in the Asia-Pacific region strive to increase their military and security significance, what are their prospects and conditions?

A number of nations have after WWII, and especially in the past decades, strived to become highly self-sufficient in arms procurement through developing an indigenous, highly sophisticated defense industry. Nations like Brazil, India, South Korea, Indonesia, Singapore, South Africa, South Korea, and Israel have been highly ambitious in this regard. Israel has been most successful in this group, albeit with very strong support from the US. Perspectives on the prospects of actually succeeding in creating strong national defense industries that are able to produce defense systems at internationally cutting-edge level vary greatly. Several nations, for example India and Brazil, have so far not come close to reaching their ambitious goals for indigenous self-sufficiency. There is a large discrepancy between the ambitions and expected socioeconomic effects compared to the actual results of most such national aspirations; there are many complex and interrelated issues that affect the outcome.[24]

Defense-industrial globalization is an uneven process. Most regional defense industrial bases lack the necessary design skills and technological expertise in order to truly innovate; they can import goods or license to assemble, produce lower-end goods—but they lack the complex, interdependent capabilities in research, industry and organization; the absorptive capacity–to be able to design and produce state of the art defense materiel, and at best these countries act as late innovators when it comes to armaments production.[25]

A decision to develop a new military capability not only rests upon the performance of the product, its price, the technology transfer, and the offset obligation setup. One important factor is also the security policy implications the defense acquisition brings with it. An acquisition of a major defense system and the creation of a new, decisive military capability constitute a security policy handshake between the seller and the buyer. The companies are situated in the middle of the implementation of the transaction, but it can never take place without the blessing of the two nations. If the buying nation foremost values to belong to the US global sphere of security interests, that will become a decisive factor. A nation might also actively strive *not* to become dependent upon the US in defense technology. For others, a dependence upon, for example, France or Russia might be seen as out of the question. The choice of defense technology thus does not solely rest upon price and performance; arguably the technology transfer and the offset obligation are more important. Security interests will underlie all factors, and at an early stage rule out certain alternative, contending defense systems.

For Asian nations striving to improve their military capabilities, but that do not possess a broad and sophisticated defense-industrial base, they could partner with sophisticated but smaller defense-industrial nations like Sweden. Sweden and the domestic defense industry has performed such partnering with Singapore (submarines and certain other naval technology areas), Thailand (fighters), and in some areas also with South Korea. If there is an indigenous defense industry, the offset obligations tend to nowadays create company-company relations that are strategically attractive for both the seller and the buyer. With the high stakes in defense export, the buyers also have a strong bargaining position towards the seller in order to create attractive technology transfer.[26]

If we compare such aspiring nations to Sweden, Sweden has overall decreased its level of self-sufficiency and also lowered its ambitions for indigenous development. Broadly, one can say that they are transforming in opposite directions, which may create windows for partnering. Finally, to comment upon the questions formulated in connection to the chapter's purpose:

What are the sources of input to national defense innovation systems? The sources come from a complex network of companies, militaries, government agencies, academia, and research institutes—domestically but also from abroad. Asia-Pacific nations can benefit from the innovation resources of the exporting nations and find synergies between mirroring organizations.

How can a nation analyze how to create a new military capability through defense materiel import? This chapter suggests a systematic approach based upon the suggestion that the nation buying defense materiel should partner with the selling nation, and aim to find synergies and nodes of collaboration with the selling nations' innovation system's focal organization (military, government, academia, and defense industry).

How can a technology policy and bilateral partnering help to improve the process of developing new military capabilities in a nation? A technology policy can be seen as one important tool for being able to define ambitions, possibilities and challenges, and strategic fit between the concerned nations in a bilateral partnering in an export-import relationship.

Conclusions and recommendations

In order to have realistic ambitions for establishing a new or clearly augmented military capability, aspiring Asia-Pacific nations must accept a technology-follower position, and not formulate unrealistic ambitions that are not matched by domestic qualities and size of its defense innovation system and of accessible financial resources. These nations should also formulate a balanced technology policy in order to define goals, ambitions, and how to achieve the goals of the new military capability. Furthermore, in order for this plan to have a long-term stability, these nations should create long-term bilateral partnerships based on knowledge and technology fit with sophisticated, exporting nation(s) and develop synergies in military, defense R&D, government, industrial, and innovation collaboration based upon the defense materiel import.

Notes

1. SIPRI arms transfer database (www.sipri.org), accessed May 2013.
2. Author's email survey with Swedish defense companies.
3. A.L. Ross, *On Military Innovation: Toward an Analytical Framework* (La Jolla, CA: University of California Institute on Global Conflict and Cooperation: 2010).
4. C. Edquist, "Systems of Innovation Approaches—Their Emergence and Characteristics," in C. Edquist, ed., *Systems of Innovation—Technologies, Institutions and organizations* (London: Pinter: 1997).
5. H. Etzkowitz, and Leydesdorff, L., "The Dynamics of Innovation: From National Systems and 'Mode 2' to a Triple Helix of University-Industry-Government Relations," *Research Policy* (2000), pp. 109–123.
6. S. Zahra, and George, G., "Absorptive Capacity: A Review, Reconceptualization and Extension," *Academy of Management Review* (2002), pp. 185–203.
7. M. Axelson, and Lundmark, M., *Industrial Effects of Direct Military Offset in Defense Materiel Export* (Stockholm: FOI: 2010).
8. M. Axelson, and Lundmark, M., *Teknologipolicy för Ledningsområde Mark— Principiella Alternativ för Realisering av Regeringens Inriktning [Technology Policy for C3I in the Land Arena—Principal Alternatives for the Realization of the Government's Directions]* (Stockholm: FOI: 2011); M. Axelson, Karlsson, C., Khan, M., and Lundmark, M., *Teknologipolicy för Ledningsområde Mark* (Stockholm: FOI: 2012).
9. M. Maidique and Patch, P., "Corporate Strategy and Technological Policy," in M. Maidique and Tushman, eds., *Readings in the Management of Innovation* (Pensacola, Florida: Ballinger: 1998).
10. Axelson and Lundmark, *Teknologipolicy för ledningsområde Mark*.
11. F. Barnaby, and ter Borg, M., *Emerging Technologies and Military Doctrine* (Houndmills: Macmillan Press: 1986).
12. Maidique and Patch, "Corporate Strategy and Technological Policy."
13. N. Slack, Chambers, S., and Johnson, R., *Operations Management* (London: Prentice Hall: 2010).
14. S. Oxenstierna, and Westerlund, F., "Arms Procurement and the Russian Defense Industry: Challenges Up to 2020," *Journal of Slavic Military Studies* (2013), pp. 1–26.
15. The decision was to upgrade 40–60 Gripen C/D to E/F together with Switzerland, which may acquire 22 Gripens. The present decision (January

2013) is that the upgrade version will only be the E version, and not the F two-seater version.

16. Under the new Social Democrat-Green Party government coalition since September 2014, there has been a clear strengthening of the Swedish military homeland defence. This has resulted in a higher priority on military readiness and increased funds to acquisition of several defense systems. The trigger for these reforms are Russia's actions in Ukraine combined with a more aggressive rhetoric and operative behavior around the Baltic.

17. Axelson, and Lundmark, *Industrial Effects of Direct Military Offset in Defense Materiel Export*; M. Lundmark, *Transatlantic Defense Industry Integration—Discourse and Action in the Organizational Field of the Defense Market* (Stockholm: Stockholm School of Economics: 2011).

18. SIPRI arms transfer database (www.sipri.org), accessed May 2013.

19. Switzerland has (based on a referendum) since chosen not to acquire Gripen. Brazil has however in 2015 acquired 36 Gripen fighters for 39 billion SEK – the biggest Swedish industrial export ever.

20. Author's email survey with Swedish defense companies.

21. Axelson and Lundmark, *Industrial Effects of Direct Military Offset in Defense Materiel Export*.

22. M. Andersen, and Katz, P., "Strategic Sourcing," *The International Journal of Logistics Management* (1998), pp. 1–13; N. Howard, and Caldwell, M., *Complex Performance* (Abingdon: Routledge: 2011).

23. M. Khan, Lundmark, M., and Hellström, J., *Sällsynta Jordartsmetaller—Betydelse För Det Försvars- och Säkerhetspolitiska Området [Rare Earth Elements—Implications for Defense and Security Policy]* (Stockholm: FOI: 2013).

24. K. Boutin, "Emerging Defense Industries: Prospects and Implications," in R. Bitzinger, ed., *The Modern Defense Industry* (Santa Barbara: Praeger Security: 2009).

25. P. Dombrowski, and Ross, A. L., "The Revolution in Military Affairs, Transformation, and the U.S. Defense Industry," in R. Bitzinger, ed., *The Modern Defense Industry* (Santa Barbara: Praeger Security: 2009), pp. 153–174; R. Bitzinger, "China's Defense Technology and Industrial Base in a Regional Context: Arms Manufacturing in Asia," *The Journal of Strategic Studies* (2012), pp. 425–450.

26. Axelson and Lundmark, *Industrial Effects of Direct Military Offset in Defense Materiel Export*.

4

Emerging Technologies' Potential to Change the Balance of Power in Asia

Eugene Gholz

Technology underlies the ways that militaries interact with each other—at its simplest, whether their forces even have the range to interact. It is by no means the only factor involved in the strategic balance. Each country's economic power has a big effect, because it determines not only what technologies it can develop, purchase, and maintain but also how much of each technology a country can afford. And a host of other factors influence countries' ability to use those technologies effectively. Perhaps even more important, foreign policy intentions, whether aggressive or aimed at preserving the status quo, are the proximate cause of many interactions. But technology and its interaction with geography constitute the structure of the strategic environment.

Yet technology is a very abstract, complex variable. In practice, militaries use a great many technologies at the same time, and there are many ways to slice technology: aircraft and ships and ground vehicles; fighters and bombers, transports and surveillance aircraft; sensors and stealth, propulsion, armor and weapons. Each weapons system—and, more, the "system of systems"—that each side brings to an interaction contains myriad particular technologies that mean that crude comparisons of the "level of technology" across militaries have very limited analytical value. And of course even within each category or technology—say, fighter aircraft—many militaries have a mix of equipment, making their general level of technology hard to evaluate. The question gets even more difficult when a military is matched against its potential adversaries, as different technologies yield different levels of relative advantage when pitted against different enemy equipment.

Nevertheless, strategists and planners, and ultimately national leaders, must constantly evaluate the effects of introducing new technologies into the strategic balance, considering whether they are worth the costs and risks of acquisition. These key officials must think about scenarios in which the technology could plausibly be used. What strategic problems would it potentially solve? This question should apply whether the technology is new only to the particular military or whether it is a research and development project to push the global state of the art, whether a relatively low-risk project to import a type of equipment widely used in other militaries or a higher-risk effort aimed at creating an emerging technology that has never been seen before. Technologies affect specific strategic balances, often in difficult-to-predict ways, rather than a general scoring system that determines which country is "ahead" and which is "behind."

However, we can reasonably ask about the general tendencies that might follow from introducing a new technology into a regional strategic balance without getting into the overwhelming complexity of the many interactions among systems and the multitude of different scenarios that each military could face. Specifically, does the new technology make it easier for aggressors or defenders? Would equipment with the new technology be more useful in a first strike or when absorbing and retaliating against a first strike? While even the most "defensive" system can be put to use in service of an offensive doctrine—for example, the Maginot Line in principle could have made defense of the French border require only a thinner covering force, leaving a greater proportion of the French military available for an offensive operation—it is still useful to think through the strategic effects of emerging technologies in the general framework of the offense–defense balance.

The remainder of this chapter will briefly consider the effect of emerging technologies on the balance of power in East Asia. It principally offers a framework for thinking about the problem. It abstracts from the complexity of scenarios and detailed models of potential combat— the domain of actual force planners, who use the tools of campaign analysis. It also stays away from detailed futurology of the particular emerging technologies that are often discussed in the region, which range from things that are already used elsewhere in the world (e.g., air-independent propulsion for submarines) to speculative developmental projects (e.g., hypersonics, which have been demonstrated in some advanced concept equipment but not really in deployed forces, or quantum computing, which may not even be that far along yet). Instead, the chapter offers a thumbnail sketch of the structural characteristics of the

East Asian balance of power and then explains three ways that emerging technologies could shift the strategic balance. Each of these ways bears further, detailed investigation, both by regional militaries and by independent strategic analysts.

A high-level analysis of the East Asian balance of power

Much discussion of the strategic situation in East Asia speculates and analyzes the intentions of key countries. Does China have revisionist aims? Might Japan? Is North Korea's leadership stable enough that we can rely on our understanding of North Korean interests? Stable both in the sense of "can the regime survive" and also, if the regime survives, are its leaders mentally stable in the sense of other countries' view of rationality? Important questions, to be sure, and the answers might determine among other things how much we should expect various countries in East Asia to invest in emerging technologies. But analysts also should put debates about intentions in the context of serious thought about capabilities: for example, if China or Japan were revisionist, what could it do to act on those intentions? Technology and geography constrain all powers in East Asia.

In East Asia today, defensive advantages create "slack" in the system. The strategic balance is not like a tautly pulled string, where just a little more pressure could stretch the string to the breaking point, giving one country a clear first strike threat against others. Forces on the operational and tactical offensive usually (though not always) have to be substantially larger than defending forces to have reasonable confidence of success, meaning that countries can often adequately train and equip for defense without posing an obvious threat to their neighbors.

Several factors create this slack in contemporary East Asia. Geographic distances between potential adversaries are in many cases relatively long, which compounds the challenges for the offense. The logistics effort to keep an effective force in the field increases non-linearly with distance. Moreover, range limits often bind more tightly on attacking forces than on defenders, because attackers use up a significant fraction of their fuel marshaling for an attack—for example, as aircraft loiter after takeoff while waiting for the other components of a strike package to assemble in the air. Of course clever attackers will strive for technical and organizational solutions to these challenges, but they are challenges nevertheless, especially facing the friction of real-world operations rather than the idealized performance of a tabletop military exercise.

Most important, many of the significant boundaries in East Asia are maritime boundaries, meaning that potential aggressors would have to overcome "the stopping power of water." Water requires special transportation to cross, makes supply for a continuing campaign even more complex and less reliable, and powerfully limits the possibilities for cover and concealment to protect an attacking force during the crossing. Japan and the Philippines have relatively large moats between themselves and plausible adversaries, and even the Taiwan Strait poses major operational challenges for an aggressor. Some other countries in the region (e.g., South Korea, Singapore) do not benefit as much from water boundaries, but other geographic features like mountain ranges aid in potential defense. Overall, although questions about intentions in East Asia make analysts nervous about conflict there, the strategic geography should mitigate those fears.

The second key factor in the contemporary balance of power in East Asia is the role of the United States. The United States spends vastly more on its military than any country in East Asia, giving the United States a large, well-trained military equipped with technologically advanced, high-quality weapons. The high level of spending, combined with alliances with a number of regional powers that provide local bases for some US forces, allows the United States to overcome the Pacific Ocean "moat" to participate in the regional balance of power—beyond the American role in extending nuclear deterrence. The US friends and allies tend to have smaller forces than their potential adversaries, whether because the countries are smaller (e.g., Singapore) or because they choose to spend less than they could afford on their militaries (e.g., Japan). The result is that the US presence provides an important backstop to the relatively weaker powers and perhaps adds to the strategic slack in the region.

So long as the United States is viewed as a benevolent external player that is not interested in aggression, the US contribution to the regional balance can compensate for a potential surge in investment by a regional military. On one hand, a weaker US ally whose defense budget spiked could still not generate enough power to launch an offensive on its own. On the other hand, a country not allied with the United States would be unlikely to gain a decisive advantage through a surge in investment because the US could shift global forces to offset the effect of the increased regional investment. The idea behind the American presence is to serve as a regional pacifier, although this idealized picture may not play out in practice. But even if not all governments in the region see the United States as a moderating influence on other locals' potentially aggressive intentions, the fact that US forces generally side with weaker countries tends to push the distribution of power in East Asia toward balance rather than out of balance.

This high-level picture of the balance of power in East Asia is not static. While changing military technology has some effect (as will be discussed in detail below), the most important dynamics in the region are China's continuing economic growth, which means an increase in latent military power, and China's expanding effort to convert economic power into deployed military forces through rapid defense budget increases. The United States describes its announced intention to add to its regular deployments in Asia (the "rebalance" or "pivot") as an effort to maintain the balance of power in the face of these dynamics.

The budget situation on both sides of the Pacific injects some uncertainty into the East Asian balance of power. If China's economy slows down for a sustained period, will it maintain its trajectory of defense spending? In the face of pressure to spend money on other priorities, not least on reducing government debt, will the United States choose to spend enough on defense to keep pace with its current strategy for maintaining the balance of power in East Asia? The larger countries' underlying economic and political choices—including, in some cases, their choices about how much to invest in emerging technologies—will be the main drivers of change in the strategic balance. But even if economic growth and defense budgets are likely to have greater effects than changes in available military technology, few plausible adjustments to defense budgets are likely to get the East Asian strategic balance out of its "slack zone:" that is, even the factors that can have the biggest effect on the distribution of power are unlikely to move it far enough to create dangerous imbalances and opportunities for aggression, given the underlying regional geography and baseline defensive advantage.

How emerging technologies could shift the strategic balance

Militaries constantly look to create advantages and to repair weaknesses vis-à-vis potential adversaries. New technologies may help them with specific missions in particular planning scenarios of interest, and they appropriately make investment choices based on such detailed analysis. But we can also consider bigger-picture effects of new technology on the balance of power, because technology can change the ways in which geography and military spending affect the balance of power. In contemporary East Asia, emerging technologies seem most likely to alter the strategic balance through three paths: by changing the offense–defense balance; by changing the fixed costs of military investment, which would disproportionately affect smaller countries; and by changing the cost of US power projection into the region.

Technology determines the way that militaries interact with geography. For example, the types of sensors available (from naked eyes through spyglasses to radars) determine what constitutes cover and concealment for advancing forces. In a marine environment, radars that can deal with the noise introduced by wave action are more effective than those that cannot, and a completely different type of sensor that tracks targets using magnetism, based on the anomaly of finding a large metal object in the middle of the sea, can have obvious advantages. A broad range of technological changes, likely focused on computing power for advanced signal processing, threaten the long tradition of surreptitious activity at sea. This trajectory directs investment in emerging technologies towards countermeasures. For example, if one of the most likely ways to detect a diesel submarine is the radar signature of the snorkel that takes air into its engine, a technological advance to allow air-independent propulsion might significantly help preserve the submarine's potential for concealment, despite the general trend that makes such concealment increasingly difficult.

But to understand the effect of an emerging technology on the strategic balance, it is not sufficient to consider how the technology would solve an important tactical problem. Analysts need to press at least one major step further: would greater concealment for submarines tend to make it easier or harder to execute aggressive operations? If the strategic challenge is to project power across a moat, a submarine might be mostly a defensive weapon: a submarine cannot carry significant ground forces or land them on an invasion beach, but it can interdict an adversary's transport and supply ships. On the other hand, maintaining submarines' concealment might help with aggressive missions like infiltrating special operations forces or laying mines as part of a blockade. The countries in East Asia that are considering investing in air-independent propulsion, distinct from the countries that already have deployed submarines with this technology, are more likely to intend to use the submarines primarily for strategic and operational defense, and the emerging technology may make their task relatively easier.

Analysts should think through the potential effects of other emerging technologies in similar fashion. The United States and perhaps a couple of other countries have enjoyed a significant military advantage in recent decades because of their ability to execute precision strikes, essentially putting fixed defenses at risk of a disarming first strike. That capability is already diffusing to other modern militaries, including in East Asia. However, defenders have maintained a significant advantage by deploying mobile launchers, whether for retaliatory missiles (like the Iraqi Scuds of the Gulf War) or for anti-ship cruise missiles in coastal defenses: the ability to precisely hit a target has had limited utility when the targets are difficult to find and have time to move out of the way between their detection and the moment that an attacker can put ordnance on the target. If emerging technologies allow for improved performance of long-range precision strike against mobile targets, those new technologies may significantly help offensive forces relative to defenders in East Asia.

The reduced time between detection and attack has been a key part of the attraction of hypersonic systems, at least in theoretical discussions, since actual deployment of hypersonic systems is further off in the future. But time-to-target may not be the most important constraint on attackers' ability to destroy mobile defenses. Instead, weaknesses in surveillance systems and especially in command and control—weaknesses that must be addressed through training and organization at least as much as through technology improvement—may be the most important constraints on potential attackers. If so, the emerging hypersonic technology may not make much difference in the East Asian offense–defense balance, although this particular tactical application is not the only one in which advocates see hypersonics as relevant to future military operations.

Emerging technologies may also change the strategic balance through a second path, by affecting which countries are relevant to the strategic competition in East Asia. Historically, relatively small countries have sometimes been able to make a real military contribution, often by acquiring small but highly trained, well-equipped forces. At other times, of course, the types of technologies available or the diffusion of technologies to both sides of a conflict have meant that large countries were able to completely overwhelm smaller ones. Technology has an important influence on whether elite forces can dominate mass forces. Sometimes that technological edge is within reach for small countries that are willing to spend a significant amount per soldier; at other times, the potential technological edge either does not exist or is too expensive for even wealthy small countries to afford.

In East Asia, the question is whether emerging technologies will raise or lower the cost for the several countries that have small, high-end military forces—and if emerging technologies raise the cost, will it price those small militaries out of the competition? At the very top end of global defense investment, defense reformers in the United States have warned for several decades about Norm Augustine's famous projection that the current cost-capability trajectory of defense investment means that even the US Air Force will only be able to afford a single airplane to equip itself in the relatively near future. While military leaders salivate over the cost savings that they observe miniaturization enabling in commercial markets, they have thus far mostly been stymied in their efforts to use complex electronics and advanced materials to reduce the cost of highly capable weapons systems. Instead, the unit cost of the electronics and materials that are so important in contemporary defense equipment persistently increases. The rising costs have already made it much more difficult for smaller militaries around the world to buy high-end American equipment in meaningful numbers; several have opted for more affordable equipment from Europe or Russia. The ongoing debates about international participation in the F-35 program put a fine point on the cost of buying the very best (assuming that the F-35 really turns out to be as capable as everyone hopes, after it comes through operational test and evaluation and the subsequent production adjustments).

Again, though, the overall effect of this cost trend may have only a limited effect on the strategic balance. The larger and wealthier countries in East Asia have the luxury of making choices about whether to keep up with emerging technologies: for China, Japan, South Korea, or Taiwan, the question is how much of their wealth they choose to spend on defense, and if they choose to spend more, they can almost certainly buy the level of technology that they wish. That may not be true for other countries that currently play a role in the East Asian strategic balance. For example, Singapore is very wealthy on a per capita basis but not as much in absolute terms; conversely, Indonesia has a large population but not as much disposable income that it might choose to divert to defense.

But to understand the implications of pricing small countries out of the balance of power, analysts must consider what those countries currently contribute. Singapore, for example, is not likely to put up a fight by itself against many other countries in the region. Instead, Singapore and other countries may buy high-end equipment mostly to signal that they care about defense and are eager to continue participating in alliances. Perhaps these countries buy what they do to show that their commitments to offer strategically valuable bases to their allies are not just "cheap talk." Perhaps their role in the alliances is already to provide moral support and diplomatic cover rather than combat power, so emerging technologies that are too expensive for them to buy really make little difference to their marginal contribution to actual fighting power in the region. These are judgments best left to regional experts, both military and civilian, but the bottom line for analysts of emerging technology is to understand that the effect of technology depends on the expected roles and missions that make up the current balance of power.

Finally, emerging technologies may have their most profound effect on the East Asian strategic balance by changing the role of the United States in the region. In recent years, the US military has devoted a lot of attention to Chinese Anti-Access/Area-Denial investments: the development of new technologies, deployment of new forces, and training of more-capable units that raise the potential threat faced by US ships and aircraft operating in the region. So far, the US response has been to spend a lot of money to push technology into the teeth of the Chinese defensive advantage. If the Chinese deploy large numbers of more sophisticated anti-ship cruise missiles, then the United States needs to replace its Aegis defense system with the more-capable Cooperative Engagement Capability; if the Chinese are making progress on an anti-ship ballistic missile system, then the United States must make progress on an anti-ballistic missile system to defend ships. But this qualitative arms race dynamic may stop at some point, because the technical solutions are getting more expensive. The technology of defending coasts is getting better, and those improvements are much less expensive than the counterpart defenses on ships and aircraft. The United States has been able to project power across the Pacific in the past because of its prodigious spending and technological edge, but the amount of power that it can project there may dwindle, even if its defense budget stays high.

On the other hand, those same technologies that threaten to keep the United States out of the region also make it harder for aggressors within the theater. If emerging technologies like anti-ship ballistic missiles or hypersonic anti-ship cruise missiles are available across the region, then defenses should get stronger everywhere—as long as regional militaries adjust their spending appropriately. One possible response to diminished American power projection would be for regional militaries to try to make up the difference, but that effort would again fly into the teeth of the increasing costs of power projection. It may make more sense instead for regional militaries to work with the growing defensive advantage rather than trying to fight against it: they, too, can buy emerging technologies that make them more effective porcupines, keeping potential aggressors away from their coasts.

Conclusion: Technology's role in military affairs

Ultimately, technology is a tool of strategy, and governments choose which technologies to buy. Technology does not "emerge" by itself. The challenge in East Asia—as in most parts of the world—is to consider how to make military technology most useful for the missions contemplated in the national strategy. The balance of power in Asia, especially as a result of the region's geography, is relatively forgiving across a wide range of possible investments. Individual countries can choose from a wide menu when they consider how much to spend and what to spend their money on. To understand which technologies are most likely to make a difference, analysts should carefully consider their likely effects on the relative cost of offensive versus defensive operations, on the ability of various countries to continue to make a meaningful contribution to fighting power, and on the ability of further-away powers (like the United States but potentially also Japan) to project power into a local fight.

5
Offsetting the Impacts of Emerging Critical Technologies

Virginia B. Watson

The confluence of three developments harbor conditions that provide opportunities for Asia's less advanced militaries to offset the technological advantages of its more emerging-technology (ET) enabled regional counterparts in the decades ahead: robust economic growth, geopolitical currents, and the accelerated pace of technological progress and proliferation. Together they cultivate a robust technology-security nexus, a key dimension defining the region's strategic landscape. The momentum for global technological growth is shifting, albeit slowly, to the region. A key factor is that for the first time in over two centuries, a majority of the world's economic growth is taking place in developing Asia led by China, India, and other emerging markets in the region, and enabling its countries to share, if not challenge, the science and technology (S&T) leadership long held by the United States and other developed nations. The global map of S&T is beginning to reflect this shift: the Science and Engineering (S&E) Indicators published by the US National Science Board show the rapid growth of global knowledge-intensive activities among Asian economies. The data points to a significant increase in the global share percentage of scientific and technical output (measured by publications and patents) from China, South Korea, Singapore, and Taiwan. In the past decade, Asian countries have topped global research and development (R&D) expenditures, with China accounting for 25 percent of the total growth. A second factor is the rise of innovation capacity among the developing countries in the region indicates that China, India, Malaysia, Mongolia, and Vietnam now belong to a select group of 18 emerging and middle-income economies rapidly improving their innovation capabilities and moving up the GII rankings, while Hong Kong, Singapore, and South Korea continue to hold their top positions. Furthermore, there is now an unparalleled expansion of technology-transfer opportunities through defense investments spurred on by burgeoning regional security needs and concerns.

Third, a cluster of geopolitical conditions—the escalation of maritime territorial disputes, the strategic implications of the rise of India and China, and the US rebalance to Asia strategy—are fueling the build-up of defense capabilities in the region. So while worldwide military spending in recent years have dipped primarily due to the defense budget cuts in the United States, Asia's military expenditures have increased, according to a 2013 report published by the Stockholm International Peace Research Institute (SIPRI). BRIC nations such as India and China have doubled or even tripled their defense spending in the last 20 years. For the first time in 15 years, Russia's military expenses as measured against its GDP exceed that of the United States. Defense spending among emerging economies in Southeast Asia—led by Vietnam, the Philippines, and Indonesia—have collectively increased by 5 percent. And fourth is the increasing strategic importance of non-traditional security issues. Climate change, humanitarian assistance/disaster relief, health, and cyber challenges among others have introduced a corresponding high table policy presence of S&T as part of national comprehensive security responses to the challenges stemming from these non-traditional issues. For instance, the defense cooperation initiatives of India and Indonesia across a broad range of technologies reflect, among other things, a rising demand in Asia for battlefield management that are suited for non-conventional missions such as border protection, counter-terrorism, and homeland security.

Conversely, continuous and rapid technological advances are shaping Asia's security environment. The rise of non-state actors and networks as empowered players in the security arena is partly due to the accessibility and availability of technologies. The use of simple, inexpensive technological inventions such as improvised explosive devices (IEDs) by non-state actors have brought into full relief the asymmetric features of conducting modern-day warfare, driving the imperative for military organizational and doctrinal adjustments. ETs such as unmanned aerial vehicles (UAVs, more commonly known as drones) and additive manufacturing (AM or 3D printing) not only offer militaries and non-state actors the possibility of creating and/or destroying competencies, but also expand the space for technological disruptions and surprises that can either stabilize or destabilize the region—game-changers, in short. The information technology revolution—while enhancing security—is also creating new forms of insecurities and exposing serious challenges in the construction of an international regime for cyber operations. Rogue or weak states in possession of advanced weaponry have used these technologies to generate strategic advantage and simultaneously foment unstable environments. For the United States, technological superiority is a vital currency of power projection, and its dominance in defense technologies has lent credence to its enduring and significant presence in Asia. Meanwhile, emerging powers India and China, poised to challenge the US position in the region, are not surprisingly building advanced technological capabilities as an important dimension for enhancing their strategic positions and their ability to shape regional security dynamics.

This technology-security nexus manifests three modalities that create offset opportunities for less advanced regional militaries: defense capacity building, the dynamics of technological evolution, and doctrinal developments.

Capacity building

Nations where the military is most likely able to develop offset capabilities vis-à-vis its more advanced, ET-enabled counterparts in the region are themselves expected to see reasonable technological progress in the decades ahead. In order to build defense-technology capacity, emerging economies are balancing the acquisition of materiel and technological know-how with the acquisition of new institutional arrangements and relationships. So while Southeast Asia is now second to India as the world's second largest importer of military equipment and technology, its nations are also stipulating local manufacturing participation in defense procurement contracts with other countries or foreign defense firms. For instance, ThalesRaytheonSystems (TRS) recently inked a $164 million deal with Indonesia to supply the country with its ForceSHIELD Short-Range Air Defense System. Part of the contract stipulates that TRS is to work with and transfer radar manufacturing skills and knowledge to PT LEN Industri, an Indonesian state-owned electronics firm. In the same vein, Malaysia's shipbuilding-to-weaponry company, Boustead Heavy Industries Corporation, is partnering with DCNS, a French state-controlled naval contractor, to locally build six coastal combat ships worth $2.8B for Malaysia's navy as a way to develop domestic R&D capability.

Key virtues of this two-pronged 'leapfrog-and-learn' strategy are that first, it lays the groundwork for the development of an indigenous R&D system using advanced technological platforms as start-off points. And second, this strategy provides, in the long run, an innovation space for less advanced militaries to develop potential counter-technologies or create new thresholds for technology use that can undermine or neutralize the existing advantages of ET-enabled counterparts. As more militaries in the region embark on defense and indigenous technology capacity-building in an environment of rapid technology and knowledge transfer, it will be increasingly challenging for more-advanced, ET-enabled counterparts to anticipate technological countermeasures indigenous to less advanced counterparts in the region, to monitor and respond to new uses of technology originating elsewhere in the region, and to offset technologies that are not interoperable to their systems and platforms.

Second, niche strategy provides a calculated opportunity for less advanced militaries to achieve technological leadership or superiority within the constraints imposed by limited resources. In this case, targeted technology capacity-building is an offset strategy, a direct challenge aimed at achieving parity or superiority vis-à-vis the technological leaders. Apart from reaping the obvious economic rewards, adopting a niche strategy can provide multiple gains for a nation's innovation system. It can offset some of the country's technological dependencies, generate competitive advantage in one or few select technologies, while simultaneously creating a network of external partners for superior, indigenous technologies. These core competency gains extend to the nation's defense technology base that perforce becomes privileged as a first-user rather than a follower in the niche areas, producing significant strategic capital for the nation. Writ large, technology niches can elevate the quality of the nation's innovation system demonstrated through, among others, enhanced system integration capability, local and global networking, rapid innovation, and tight linkages between the defense and civilian sectors.

Asia is not lacking in examples of emerging economies that have been successful in employing the niche strategy. Taiwan and Korea—both late entrants in the thin film transistor-liquid crystal display (TFT-LCD) industry—successfully employed it to upend then global leader Japan to subsequently dominate the market. Today, Taiwan controls the world's small- and medium-sized LCD panel market, while South Korea holds a major share of the large-sized LCD panel market. An increasing number of developing Asian countries are employing the niche strategy and investing in ETs: forecasts indicate that—excluding Japan—the largest increase in spending for UAV R&D by 2022 will be in China, South Korea, Indonesia, and Thailand. And while the US currently leads in

3D printing patents, Singapore and China are poised to surpass its R&D investments in the near term. The rapid pace of technology transfer will accelerate niche development in the region among its emerging economies; it is one clear path to offset, if not challenge, the advantages that advanced, ET-enabled militaries acquire from advanced, spin-off or spin-on technologies.

Finally, the rise in the strategic significance of non-traditional security issues is a modifier of defense niche capacity-building. There is a growing recognition among defense and national security policy makers in the region that the convergence of traditional and non-traditional challenges will be a permanent feature of the future global security environment. Response operations to hybrid challenges such as HADR, countering violent extremism (CVE), and cyber threats perforce require extensive cooperation across nations and between government (civilian and military sectors) and non-government actors (private sector, NGOs) on both national and international levels. This fluid environment of exchange is fertile ground for generating disruptive and/or offset innovations because the knowledge sharing and technology transfer across nations (and their militaries) as well as between the public and private sectors is expediting the process of modern technology acquisition among developing economies. For instance, the increasingly widespread use of drones and IT for HADR and CVE operations in Asia have provided more opportunities for less advanced militaries to upgrade their skills and knowledge, and in some cases, acquire sufficient know-how to develop indigenous versions of emerging technologies (e.g., Vietnam drones). So while the transnationalization of security threats will lead to greater international technology cooperation and capacity-building, it will also create increased possibilities for offset innovations to occur in emerging economies as they acquire or adopt advanced technological systems. Environmental issues such as climate change and natural resource availability will be key parameters shaping regional security dynamics in the decades ahead. This cluster of non-traditional security issues requires not only "business-as-usual" technology solution sets, but necessitate the accommodation of a "sustainable" approach to technological development. Sustainability is a growing strategic platform in Asia. For one, it is the region with the largest energy demand and lowest fossil energy resources, and more than half of the world's population. In addition, data projections show that Asia will be the site of substantial green growth as global leadership in renewable resource markets shifts towards the region's developing economies. In this context, national systems of innovation are expected to evolve towards integrating a sustainable feature as resources become scarce and manifestations of climate change become more commonplace. Paradigmatic adjustments are on the way: recognizing the growing importance of the green growth-innovation nexus, the annual GII in 2012 included, for the first time, "ecological sustainability" as a key pillar of national innovation input. Inclusion of this criteria reconfigures the innovation positions of countries, with some advanced nations going down the ranks while other developing ones moving up the index.

The role of the military is an integral part of this wider evolution in the concept of sustainability. The creation of a sustainable green defense, or at least the integration of sustainable features in the development of military capability, is an innovation opportunity that creates long-term strategic benefits for its adaptors. So it is no surprise that climate change and the attendant environmental issues are articulated in the latest Quadrennial Defense Review—the US Department of Defense (DoD) assessment of its strategies and priorities—and also included in the US State Department's policy agenda and regional initiatives. DoD's Green Procurement Program (GPP) Strategy and the recent announcement of the US Navy's intention to "sail the Great Green Fleet of 2016" that would signal the commencement of the Navy's "new normal" underscore the US defense' explicit recognition of sustainability as a strategic paradigm of the future. So an argument can be made that this path also provides less advanced militaries (a state institution that is one of the largest consumers of energy) as early adaptors or fast followers a platform to leapfrog resource-intensive, unsustainable technological legacies originating from the West and chart a capability trajectory that is more sustainable. Developing an early technological interface between itself, the defense industrial base, and the green S&T private sector, while not directly challenging or countering the technological advantages of its more advanced counterparts, can in due course provide their countries with a supportable, ecologically responsive defense capability. Looking ahead, it is reasonable to suggest that strategic survival, beyond looking at transformational technologies, will be a contest about alternative futures, one of which is defined by sustainability.

Technological evolution

Features of today's interconnected technological currents create conditions that increase the prospects for Asia's less advanced militaries to offset the technological advantages of more-advanced ET-enabled regional counterparts. First, technological evolution is occurring at an accelerated pace resulting in the proliferation of accessible, IT-based technologies. Bolstering this trend is the expansion of the traffic of technology and knowledge transfer to include networks of non-state actors, individuals, and corporations. In the hard security domain, these converging developments have translated into more power and strategic leverage in the cyber domain for less advanced militaries. Low barriers and cost to entry, the problem of attribution, the unpredictability of cyberattacks and crimes, and the absence of an international legal framework make IT an "equalizing" or "neutralizing" technology that is creating enormous opportunities for less advanced militaries to achieve strategic parity vis-à-vis their more superior counterparts in both defensive and offensive terms. For advanced militaries, the challenges posed by less capable and less-resourced adversaries in the cyber domain are extensive enough that simply buying or creating more and better hardware is no longer enough to maintain credible capabilities for power projection.

Second, innovation is no longer the monopoly of the developed West: where the advanced Western nations were at one time the only sources of the world's technological advancements, the geography of global innovation shows the increasing contributions of Asia's emerging economies in technological development, i.e., the growing role of India, China, and the emerging Asian economies as nodes of leading edge technological development and innovation. Furthermore, South-South' technological exchange and transfer is increasingly an integral feature of the "new normal" in technological development founded on common security concerns and approaches. This dynamic environment is giving rise to a gamut of shared technological innovations – from simple to advanced – that provide less powerful nations in the region the tools to cope with their security challenges. Technologies or technological platforms appropriate only to the needs, requirements. and competencies of developing economies may not necessarily be compatible with those from the West. The Science, Technology, and Innovation agreement between BRICS member-states (Brazil, Russia, India, China, and South Africa) cover an array of S&T areas that directly leverage the states' respective competencies, thereby promoting the development of indigenous knowledge systems. These identified areas are potentially bases for technology offset strategies because the trajectories of technological development may digress from the course set by advancements from the West. Indeed, in response to growing South-South S&T cooperation, defense institutions in advanced nations such as the U.S. and Western Europe are making institutional and structural adjustments to ensure that their technology scanning, acquisitions, and interoperability plans and policies are effectively tracking S&T developments in the developing world, finding ways to engage and collaborate, and identifying potential disruptions and surprises from these advances.

Second, the global IT revolution and the advent of both AM and drone-technology use in both the civilian and military domains illustrate the *de minimis* role of governance in emerging technologies. This is a derivative of the diminishing role of governments in technology R&D as states lose ground to the commercial sector as the principal agent for innovation. The nexus of rapid technological development and diffusion and the absence of an international regime governing ET use afford less advanced militaries innovation space to create new capabilities that can offset, if not counter, the technological dominance of ET-enabled militaries. ETs such as IT, AM and drone technology occupy 'ungoverned' spaces, introducing uncharted strategic, legal, and ethical issues that digress from traditional international agreements and conventions including those that govern modern warfare. The asymmetry between rapid technological developments and, at best, the glacial pace in building international regimes for IT and drone technology is a gap that both advanced and less advanced militaries can and will exploit to their advantage now and into the future. This is a situation where less advantaged militaries can set some terms of the strategic engagement. Enhancing cyber capability is top in the S&T priority list of economies like Malaysia, India, Taiwan, China, Russia, and South Korea because information technology underpins all advanced and ET platforms. It creates enormous opportunities for less advanced militaries to achieve strategic parity in both defensive and offensive terms.

In the case of drones, the United States aspires to preserve its preeminence in technological development and use with minimal constraints as a way to enhance its global strategic position. The absence of a clearly articulated US drone policy to guide UAV use in other parts of the world has provided the United States with a wide range of operational and tactical options not necessarily subject to existing international legal checks. At the same time, however, the absence of an international regime governing drones is also cause for serious concern to the United States because this legal vacuum makes highly likely the possibility of rapid and uncontrolled proliferation of technology developed in other countries (failing and failed states included) that can undercut the US position. Thus the existing permissive environment, while creating strategic gains for the United States, can also provide less advanced militaries with opportunities to develop both competitive drone technologies and counter-technologies to offset US leadership and animate geostrategic changes. Even technologically superior nations cannot anticipate all potential countermeasures nor precisely determine the response thresholds of weaker adversaries. This element of technological surprise creates leverage for its weaker adversaries.

Third, disruptions do not necessarily require developing militaries to achieve technological parity with more advanced counterparts. As Paul Bracken argues, the gap between the US military and that of a rising country such as China is not the gap that the disruptive technology must fill, but only the gap to reach some midway level of what China needs, while China can be way behind in other advanced technologies. For instance, the integration of old technologies like over-the-horizon radar with cruise missiles that can potentially kill anything and anywhere in the western Pacific out to 2,000 miles is sufficient to provoke enormous regional geostrategic changes.

Finally, technology offsets do not necessarily require less advanced militaries to achieve technological parity with more advanced counterparts. The proliferation and use of less sophisticated versions of mature technologies are sufficient to instigate strategic disruptions and create leverage for less powerful countries. The nuclear bomb, whose technology dates back to 1945, poses a real danger to international security and stability. Nuclear bomb development in North Korea, Pakistan, and Iran may be not quite as advanced as the United States, but their nuclear arsenal, to varying degrees, work enough to provide these countries with geostrategic leverage vis-à-vis their neighbors and the United States. While clearly outmatched by superior US military forces in a face-to-face conflict, they have the option to aim their missiles at US regional allies with the intention to destabilize the region and punish regimes that, for example, host US defense materiel or allow overfly rights.

Doctrinal developments: Southeast Asia

One cluster of geostrategic developments driving the spate of military modernization programs in Asia today includes the U.S. rebalance to the region, the rise of China, and the interstate maritime and territorial disputes in Southeast Asia and Northeast Asia. As a key pillar of President Barack Obama's rebalance policy, strengthening U.S. relationships with allies and partners including emerging powers such as India and Indonesia has translated into increased U.S. military-to-military engagement in the region. This strategic reach-out is perforce accelerating the defense modernization of U.S. allies and partners, with a focus on upgrading, if not developing new, naval and air capabilities. The emerging power-projection capabilities of China – as evidenced by developments such as the introduction of modern amphibious ships and planned aircraft carrier force – is partly shaping the procurement choices of its neighbors, generating a demand for area-denial capabilities such as submarines and land-based aircraft with anti-ship cruise missiles to enhance protection of their national territorial boundaries. China's maritime claims and rapid terra-forming in the South China Sea has provided an accelerant to the defense modernization intentions of the four Southeast Asian claimants (Brunei, Malaysia, the Philippines and Vietnam) contesting Chinese assertions. While not intended to challenge the technological superiority of rising giant China, modernization and capacity-building among the less advanced militaries in Southeast Asia is a trajectory that creates opportunities for generating, in the long-run, technological as well as strategic offsets.

National security doctrines underpin the pursuit of technology strategies that favor the development of specific capabilities. In Southeast Asia, the doctrinal core of this force modernization has two features. The first principle underpinning the modernization and expansion of defense capabilities is self-sufficiency. In Indonesia, the government's promulgation of the Defense Industry Law in 2012 articulates a legal-political pledge to develop Indonesian local defense industrial capacity away from heavy reliance on imports. One of the three fundamental principles of Malaysia's national defense policy, "Self Reliance," also conveys a similar commitment insofar as defense development is concerned. In 2013, the Philippines' Department of National Defense revived the 1970s Self-Reliant Defense Posture (SRDP) program as a component of the country's ongoing military capability upgrade by developing its domestic defense industry. SRDP is considered an integral part of the country's strategy to establish a "minimum credible defense posture." While foreign defense companies presently dominate the domestic acquisitions market in Southeast Asia and national defense spending is focused on importing sophisticated technologies and platforms to upgrade capabilities, nations in the region are setting in motion the process of constructing home-grown defense industries. Informed by increasingly well-defined and better articulated defense doctrines, this trend is expected to grow, creating a technological momentum that will undoubtedly give these smaller, less advanced militaries some leverage in their efforts to defend national security interests.

Self-reliance puts the focus of building local defense industries on appropriate technology, driven by the needs and requirements of each nation. The potential to develop technology-strategic offsets will therefore vary across the region. The militaries of Southeast Asian nations involved in the South China Sea dispute are modernizing and building capacities at a rate faster than those who are not party to the contest. It is likely that offsets will materialize among these nations first. The nature of a nation's alliance system could define the trajectory of technological development. The more diversified the sources of platforms during the process of materiel acquisition and local R&D tie-ups, the greater the likelihood of nurturing more technological breakthroughs or disruptions. As a treaty ally, the Philippines has primarily relied upon the U.S. for its defense development needs, so its technological platforms have a higher degree of interoperability with U.S. systems. In contrast, Malaysia and Vietnam have explored multiple international partners to build military capacities, and so will have a more customized, independent portfolio with a lower degree of interoperability with any one partner. This environment increases the possibility for indigenous breakthroughs that could be used for creating strategic advantage.

The second principle characterizes modernization and capacity-building as defensive in essence and scope. Southeast Asian nations portray the common goal to develop a local defense industry as defensive in nature and not a part of sub-regional arms race. The need to protect territorial sovereignty is foremost among these nations, and the desire to develop a domestic defense industry, to varying degrees, is a requisite to adequately defend national boundaries. For Southeast Asian nations, a "stability-inducing" modernization means an improvement of defenses without becoming a threat to neighbors and is justified by deterrence obligations, i.e., every nation's conventional deterrence remain effective and current. The maritime and territorial disputes in the South China Sea nevertheless is a forcing function that has compelled the weaker militaries of the Southeast Asian claimants to articulate a defense policy that serves as a basis to integrate, institutionalize, and indigenize defense technological development. The South China Sea contest has made naval systems and technologies as the preferred platforms for acquisition and domestic defense industry development, particularly for those countries involved in the dispute. The preference for this 'defensive tool' lends itself to be the one of the most probable sources for indigenous innovation and disruption to Asia's strategic environment.

Conclusions

It is only a matter of time before less advanced militaries in Asia will develop sufficient capabilities to potentially produce technologies that can offset their ET-enabled counterparts. The confluence of economic growth, geopolitical trends, and rapid global technological progress and diffusion is creating an enabling environment that is driving the accelerated pace of regional defense development. It brings to the fore the strategic importance of the technology-security nexus, a key dimension shaping the regional security environment articulated through defense capacity-building, the process of technological evolution, and doctrinal developments. As Asia assumes greater global strategic significance moving forward, so will its role in charting technological pathways.

The discourse of this chapter is centered on state-to-state dynamics, but it is clear that in today's world, technological evolution is seeing the diminution of the state's role and the expansion of space to individuals and small groups who are developing advanced technologies to articulate their definition of preferred societal goals and alternative futures, nongovernmental organizations that may or may not be collaborating with governments, transnational criminal networks capitalizing on new technologies for financial gain, and terrorist groups and other shadow organizations seeking to overthrow established regimes. So while it is completely relevant to study the technology-security nexus within the context of states, this examination is only a partial capture of the strategic landscape. Broadening the analytical scope of the study to examine, for instance, how superior and less advanced militaries and economies offset the technological advantages of transnational networks, or the extent to which nongovernmental organizations and individuals can counter the technological advancements of states, provides a more nuanced understanding of the notion of power and the changing concept of security. In addition, an exploration of the role that governance plays in technological development and use is necessary for three reasons. One is the growing recognition that the security landscape is increasingly both traditional and non-traditional in scope. In this respect, the question is the extent and nature of governance for dual-use technologies within the context of a comprehensive security framework, all the more important in light of the diminishing role of the state in technological development. A second is seeking formulations to alleviate the tension residing between national norms and values with international precepts that guide the behavior and policies of nations. And finally, addressing the challenge of ensuring that non-state actors are recognized as key players when formulating and implementing rules and regulations with respect to technological development, use, and diffusion. They are not only increasingly the source of ideas and innovations writ large but, more importantly, they demonstrate the potency of technology as a currency of power residing outside of the state.

6

Effective Absorption of Emerging Technologies in Defense Automotives to Enhance Land-Based Military Capabilities

Kogila Balakrishnan

The defense automotive sector had been much neglected in the past decade. The defense automotive sector is defined by land-based vehicles or combat vehicles consisting of main battle tanks, infantry fighting vehicles, trucks, and other variants.[1] Land-based systems were most popularly used by the military to face counter insurgency warfare mainly during the Cold War. In the aftermath of the Cold War, the defense automotive sector was downplayed at the expense of building air, sea, and security capability to face different forms of threats, posed by non-state actors.

In recent years, however, there is an increasing demand to develop emerging critical technologies (ECT) in the defense automotive sector driven by four key factors. First, not only the increasing amount of counter insurgency warfare but also the more complex and dangerous nature of this war which requires modification and innovation to the existing military platforms; secondly, pressure to upgrade land-based platforms to integrate with maritime and air platforms which are becoming increasingly sophisticated; thirdly to modernize the army skill and workforce. Fourthly, non-traditional security concerns such as energy and environment factors have also contributed to innovation in the defense automotive sector. The question is then to what extend the military has been able to enhance military capability through the emerging technologies in the defense automotive sector to face new challenges in the battlefield.

This chapter defines ECTs, discusses the key challenges facing the military in a land-based war environment, analyses the ECTs in the defense automotive sector, evaluates what the implementation challenges are in trying to innovate in the defense automotive sector, and finally looks at possible ways forward for sustainable technological innovation in the defense automotive sector that could assist land-based military capability.

What is ECT?

Emerging technology is defined as one that expands new technology in some significant way with a new technological development. Emerging technology is normally seen as being at an early stage of development and often takes time to reach the market or be commercialized. Emerging technology has brought profound advances and innovation in various fields of technology. When the technology is labelled as critical then the technology may have a profound impact on a nation's economic growth and development. In defense, ECT relates to technology that is so fundamental to national security and highly enabling of economic growth that the capability to produce this technology must be retained or developed domestically.

Each nation will have its own list of national ECTs and defines ECT according to the nation's local context and national security and economic concerns. The UK MOD Defence Concepts and Doctrine Centre's (DCDC) Strategic Trends Programme, the US National Intelligence Council Global Trend Programme, the French Ministry of Defence, and the European Defence Agency (EDA) shows that emerging technologies feature predominantly in autonomous systems and robotics, swarming autonomous micro aerial vehicles, development in nanotechnology sensors, cyber space, directed energy weapons. Other areas include advances in microsystems, nanotechnology, unmanned systems, communication and sensors, digital technology, biology and material sciences, energy power technologies, neuro-technology, cybersecurity, and cyber warfare.[2] The US government legislation has a list of national critical technologies which include energy, such as energy efficiency, energy storage and management; environmental quality; information and communication; living systems such as biotechnology and medical technologies; manufacturing including product manufacturing, continuous materials processing, nanofabrication, and machining;

materials such as ceramics, composites, polymers, and structures; and areas of transportation.[3]

In the past most ECTs had concentrated on spin-offs with a flow-down of ECTs from the defense to the civil sector such as radar technologies, ICT, and space technology. This trend has now changed with defense spending being reduced and defense R&D becoming much more expensive. ECTs in recent years have been borne in the civil sector with a spin-on into defense which is cost effective and economical. Countries such as Japan have adopted the spin-on method of transferring ECT from civil to defense decades ago and have been very successful in technology cost utilization. However, spin-offs and spin-ons happen less than expected due to a robust fire-wall and lack of trust, especially within defense. Greater flow and sharing of technologies by allowing for both spin-on and spin-off can certainly enhance the movement of ECT between sectors. At the same time certain ECTs such as in ICT which is much more pervasive and dual-use in nature have a wider application in both the defense and civil sector. This can be seen through the profound impact and influence of cyber-attack which goes beyond the civil sector such as banking into military and international relations domains such as diplomatic espionage and cyber terrorism.

Nations with access to ECT will have greater comparative advantage and economic prosperity. In the Asia-Pacific region, most governments have pledged to move up the industrial value-chain for economic prosperity which is seen as an effective method to create employment in high-technology sectors that are assumed to provide better jobs and remuneration. At the same time, there is a political push to absorb technology and enhance industrial capability as a way to close the technological divide that is claimed to be widening between the developed and developing nations. Owning ECT is also seen as having superiority over another nation and a form of power projection and sense of being independent. History has proven that rich nations automatically invest in a strong army. Richard J. Samuel in his book, *Rich Nation, Strong Army: National Security and Technological Transformation of Japan* discussed how Japan had used technology to rebuild its defense and aerospace industrial base to become a leader in the Asia-Pacific region.[4] Many claim that China seems to be building its technological base especially in ECT sectors for similar reasons in the 21st century. However, power projection need not be the only reason for acquisition and development of ECT.

Some nations want to adopt robust Science &Technology (S&T) policies to simply embrace technological modernization to become an advanced nation. Such countries include Singapore, Switzerland, Finland, and Sweden. These nations have a very strong S&T policy with high incentives for innovations and investments into ECT mainly for economic advancement. The defense acquisition strategies of some nations like the UK and the US do outline ECT in defense and the military. It is vital that these technologies are captured as part of the capability plan in procurement exercises whether the equipment and services are being purchased from overseas or indigenously developed.

The next section discusses the key challenges facing the military in the land-based war environment.

The changing landscape of land-based warfare

This chapter identifies three key challenges facing the military in a land-based war environment: first, changes in the nature of land-based warfare; second, issues related to energy security; and third, environmental security issues. The following section discusses in detail the above mentioned security challenges, how the military are vulnerable to such threats, and the requirement for emerging technologies in the defense automotive sector that could equip the military with adequate capability to face security challenges.

The end of the Cold War changed the landscape of the defense automotive industry. The reduction in counter-insurgency threats created a lower demand for land-based vehicles. Large defense firms channeled their resources to developing air, maritime, and homeland security products to face non-traditional security challenges and non-state actors such as terrorism. The past decade has been about inventions and innovations to serve this type of threat as opposed to the traditional counter-insurgency nature of war. The reduction in defense budget has motivated defense ministries to anticipate possibilities such as developing modular land formations. States have also changed their focus to concentrate military procurement on buying state-of-the-art naval equipment and fighter aircrafts, transport aircraft for humanitarian and natural disaster as well as helicopters and UAVs for coastal surveillance, maritime patrol and so on. This left armies around the world but mainly in Europe with ageing fleets of main battle tanks, trucks, and combat vehicles.

The neglect towards the land based scenario changed in recent years when civil war erupted in countries such as Iraq, Afghanistan, and the upsurge of war in North Africa as well as the Arab Spring. All of them requiring a large presence of army with land-based vehicles to face counter-insurgency threats. There is now a requirement for large number of ground troops with heavy armored vehicles as well as special-purpose vehicles, protected patrol vehicles, trucks, and people carriers being deployed in the warzones to face the real challenges on the ground. The military relies heavily on trucks, armored vehicles, and other variants to carry its troops and goods from one point to another within conflict zones. Despite the expanded use of cyber warfare, covert special-forces and drones to overcome problems, there is a real need to have ground troops on vehicles at the battle ground. The increasingly sophisticated nature of the threat and vulnerability has increased the risk of traveling on the ground for soldiers in the theatrical environment. Every day, soldiers are being deployed into harsh and aggressive environment killed due to roadside bombings and the use of more sophisticated IEDs (Improvised Explosive Devices) by terrorists. The vehicles crash onto the roadside bombs as occupants are exposed to components underneath the vehicle in a blast. The soldiers face multiple challenges such as carrying of overweight items on heavy vehicles exposed to IEDs, lack of a good braking system on their vehicles, and long distance travel with issues such as fuel efficiency. In 2011, the *Washington Post* reported that 268 US troops were killed by IEDs despite counter measures such as mine-clearing machines, fertilizers, and sniffing dogs and blimps with sophisticated spy cameras.[5]

Being an issue for most soldiers in an environment such as Afghanistan and Iraq, existing vehicles like the Land Rover or snatch vehicles were modified to meet urgent military requirements. However, it is not possible to add on protection to existing base vehicles as it is going to make the vehicle heavy, slow, and ineffective for transport duties, as well as the protection being rather limited and flawed. The changing nature of war requires the military to change their requirements and the new threats in land-based operations requires modifications to the military vehicles to face such threat and to handle the logistics to make it more cost effective. The military need vehicles that are more robust, made of special materials that can withstand the sophisticated IEDs, use light-weight materials in the production of vehicles, and develop good braking systems for the vehicles that can stop in time from crashing onto the IEDs.

The increasing importance of energy security

Another major driver for technological innovation in the automotive sector is related to the increased focus on energy security. This industry faces unprecedented challenges over the next 20 years as the world moves inexorably towards a new, low-carbon technology model. In the automotive sector, the goal is to support a global reduction of greenhouse gas emissions of 80 percent by 2050, The Automotive Council has set target of 50 percent weight reduction of cars by 2030. The rail industry aims to reduce CO_2 emissions by 34 percent by 2020.[6]

There have been many debates in recent years as to the increasing demand for energy, especially crude oil and gas with the rapid level of industrialization, particularly in emerging new economies such as India, China, and Brazil. At the same time, it is predicted that energy resources will diminish in the next 40 years. Although the price of crude oil had depreciated drastically in recent years especially with the availability of alternative sources of energy such as shale oil this is seen as a short-term symptom. It is predicted that there will be a continuous demand for energy sources due to the rapid pace of industrialization. Research had been undertaken to constantly find alternative sources of energy but also ways to make the auto industry more fuel-efficient.

Energy is today looked at by the military as a capability. It is important to analyze how energy used on military platforms can be fuel-efficient to make vehicles operating in the battlefield lighter but also to reduce the logistic costs of transporting fuel to faraway places. The reduced logistical handling of fuel subsequently also reduces manpower needed to handle them, and that has a direct impact on the number of soldiers that will have to be exposed to dangerous warzones. Another important constituent of energy is as to how can energy be stored effectively for operational bases. Soldiers have to operate in isolated areas without any access to alternative energy. Equipment such as hand-held radios, mounted helmets, and infrared cameras require batteries. How can soldiers be equipped with equipment that has effective energy storage capabilities in order for them to carry less spare batteries and travel light? The UK MOD for example has highlighted the importance of energy and building energy capability in the UK National Security Strategy,[7] National Security through Technology[8] document, and the UK Defence Technology Strategy.[9]

Reduction of carbon footprint and "going green"

The third factor that drives change in the automotive sector is related to environmental security issues. The increasing number of natural disasters is said to be affected by the damage on the environment including the very high levels of gas emissions by vehicles. This problem will grow as the number of cars and other vehicles increases and as more people around the world become able to afford to use cars as their main mode of transport. There are campaigns to reduce the carbon footprint and go green. Green technology initiatives have been introduced in vehicle manufacturing. At the same time, ineffective and low-efficient vehicles with greater carbon emission will increase fuel consumption. In the defense sector, there are many aging vehicles with high levels of gas emissions operating in warzones thus causing huge increases in fuel consumption. Requirements have been established to create technology that could control the level of carbon emission and make vehicles more effective with less fuel consumption for reasons of cost efficiency, reduced vehicle weight, and an effective lean logistics support.

There are several emerging technologies that were developed in the automotive sector to address some of the issues above. The technology could have originated in the civil automotive sector which then has been modified to meet the military demand. This is because there has been more funding channeled into the civil automotive sector in recent years to meet the demands of customers wanting better quality and more sophisticated cars with multiple functions at a reasonable price. The auto industry has also become highly competitive since emerging markets such as China, India, and Brazil entered this industry.

ECTs in the defense automotive industry

Defense vehicles vary from lightweight vehicles such as trucks to wheeled and tracked armored personnel carriers. Military vehicles normally serve as troop carriers, armored reconnaissance vehicle, armored assault vehicle, armament carrier, and mobile command and control unit. In most countries, the defense automotive industry will form a subset of the automotive industry. These vehicles will also generally be fitted with artillery and other weaponry for protection. The history of technology advances in defense land-based vehicles go a long way back as major wars in the past were focused on defending counter-insurgency threats. While the civil automotive industry consist of high volume, quick turnover, and returns on investments, the defense automotive sector is generally low volume, of greater complexity, requires huge capital as well as investment, and long lead time for returns on investments.

Two key observations are, firstly, the ECT in the defense automotive sector is largely focused on adaptation to existing equipment and services to meet military requirement rather than "disruptive innovation"[10] for creation of new equipment. Second, the reduced defense budget has encouraged the military to embrace an open innovation policy[11] in developing ECT by tapping into the civil automotive sector, such as the auto industry, which has allocated much more research and development fund for innovation leading to commercialization.

Some of the newer findings and technological innovation have focused predominantly on the civil automotive sectors. Most of these technologies are already applied in the commercial automotive industry and others are now being extended to the military environment in line with the military quest for innovation, reduced costs, and reduced acquisition cycle-line. The motorsport industry for instance is seen as having the ability to contribute its expertise for cutting-edge defense technology growth.[12] There are still many underlying issues on civil–military cross-fertilization and mobility of technology but greater demand for effective and efficient spending has pushed for such innovation.

Lightweight materials

One such innovation is light weighting. Research was undertaken to find ways to reduce the weight of vehicles by using light weight material to minimize metal work underneath the vehicle especially for vehicles exposed to IEDs. New material that improves protection and incorporated in more specific vehicle structure designs has greater success in achieving good levels of both protection and utility. The use of materials such as composite monocoque to shell the crew with a complimentary armor pack and V-shaped hull to deflect blast energy under the occupants. Power train parts together with front and rear axles are mounted in modules fitted to either end of the monocoque. These are designed to sheer-off in the event of an IED detonating underneath them, reducing the transfer of blasts energy and reaction forces to the crew compartment. The usage of such special material to reduce weight has also increased vehicle mobility agility, payload, and speed, thus providing for greater flexibility of operations and wide range of terrain improved transportability.[13] Another development to provide greater protection to the vehicles has been the replacement of steel with high strength aluminum alloy and titanium. Another innovation is the windscreen simulation model using computer aided engineering (CAE) that is much more efficient and has accurate solutions to protect occupants from wind screen blast dangers in war zones. These are incremental technological innovations as add-ons to existing vehicles rather than radical or disruptive innovation.

Energy storage and management

Another area of technology innovation in the automotive sector has been energy storage technologies that consider cost, reliability, weight, volume, maintainability, environmental performance, and overall conversion efficiency. Batteries are critical enabling technologies for development of advanced, fuel-efficient light and heavy duty vehicles—to promote diverse supply and delivery of reliable, affordable and environmentally sound vehicles. There has been plenty of research done to find effective energy storage methods such as research into long life of batteries and power packs for these purposes.

Energy storage research has been extremely vital in the commercial sector but there is now a huge demand from ministries of defense to assist in improved energy efficiency and to extend this technology to the military for improved energy storage technologies for operational bases. Research is focused on finding ways to increase battery life longer and store them effectively so that military can use them for longer time in isolated places. In the UK, for example, energy also forms part of the MOD Treasury Sustainable Reporting Guidance.[14] The military needs to understand energy effectiveness and efficiency of different capability options and how the energy consumption is related to military tasks. Energy storage management methods are important for the military especially if the soldiers are carrying handheld radios and operating in an environment where they will not have alternative sources of energy. The military is working very closely with the commercial automotive sector in understanding energy storage and management to enhance their capability.

Additive Layered Manufacturing technology

An emerging technology that is said to be very useful for the automotive sector is additive layered manufacturing and the creation of 3D printing technology. 3D printing machines work by using a digital description of an object to build it in physical form, layer by layer. 3D printing parts are said to save materials and only materials that are needed to shape the part are used. The printed parts can be made lighter than forged parts and the process saves fuel. 3D printing is good for prototypes of parts because it is cheaper and more flexible than tooling. 3D printing is increasingly seen as a useful tool in the automotive sector by manufacturers as an alternative to cutting, bending, pressing, and molding.[15]

Hybrid-energy technology

In terms of energy saving and reduced fuel consumption, there has increasing research into creating hybrid-energy cars and busses. The research into hybrid-energy technology has been in existence for some time but this research has become much more important with increasing attention paid to alternative energy resources and the need to reduce carbon emission. However, hybrid cars are still not as popular and are clouded by various short falls such as battery life and the frequency and length of time needed to charge the vehicle. There is still a lot of research being done to make the hybrid vehicle not only more effective but more affordable. The military is gradually venturing into this technology and is closely working with the commercial industry to test and trial hybrid vehicles as a source of capability. BAE Systems, for example, had developed the first hybrid combat vehicle which is supposed to be faster, quieter, and more fuel-efficient than the standard 70-ton combat vehicles.

Low-carbon-emission vehicles

Transportation of today is unsustainable in terms of energy usage, impact, efficiency, and cost effectiveness. For example, transport generates 25 percent of UK carbon emission.[16] Technology solutions have been developed to address the high level of carbon emissions and pollution. In the UK, a collaborative initiative between leading automotive companies and research partners was aimed at changing the way low carbon vehicles, including full battery vehicles and hybrid vehicles, are designed and developed in order to significantly reduce carbon emissions. A £29 million project was funded by Advantage West Midlands, the European Regional Development Fund, together with an industry collaborative effort between UK OEMs, consultancies, suppliers, and academic institutions into a collaborative program to create the required R&D capability and capacity for the development of key low and ultra-low carbon vehicle technology. The collaboration was between Coventry University, MIRA, Tata Motors European Technical Centre, and Jaguar Land Rover, WMG, Ricardo, and Zytek Automotive. The research concentrated on 17 different areas with each project having a lead or champion and a project team from another organization.[17]

Implementation challenges

Meeting the standards for urgent operational requirements (UORs)

Several issues had been at the forefront of ensuring that the new and emerging technologies in the automotive industry are able to flow through to the military. In developed countries, in an environment such as in the UK, where most defense equipment is now being purchased to meet urgent operational requirements (UORs), it has become difficult to respond to technology requirements. Existing products are being modified but these products are not able to meet safety and security level for military standards.

Lack of civil-military integration

Defense and civil industry have also had their differences in technology sharing. Both sectors still work in silo negating the aspiration for increased spin-on and spin-off effect. Despite the increasing research and technological innovation in the commercial sector, transfer of this technology know-how into the defense automotive sector is still minimal. Cross fertilization is more rampant at the lower level of technological readiness as compared to the higher level when it gets much more complex and difficult. Further, despite the open innovation policy introduced by governments especially in defense, the military is still suspicious of technology sharing. The military is also worried about technology seeping through commercial to other (rogue) countries.

Affordability to embrace emerging technology

In developing countries, the issue is completely different. It is a question of being able to afford these emerging technologies as well as setting priorities to acquire and incorporate such technologies into the military system. Developing countries are mostly at a lower point in the technology curve and are often trying to catch up to close the technological gap. For countries in the developing world, energy and environmental issues are often neglected and take a lower priority than other developmental issues. Further, some of these developing countries with abundance of energy resources are less concerned about energy depletion. For other poorer countries, the issues are about being able to afford third generation vehicles that are still operational and these countries still depend on soft loans and aid to build defense capability as opposed to embracing state-of-the-art technology. There is a lack of resources to embrace these technologies and to embed them within their capability environment. Also, the military and industries in these countries lack the awareness and skills needed to develop and maintain equipment with such capabilities. Most importantly, the political will is divided between providing the military with new technology as opposed to addressing fundamental socioeconomic issues. The issue for such countries is when their army is not equipped with similar technologies and standards while on joint exercises on the battlefield or on peace-keeping missions where soldiers are exposed to similar or equal levels of danger. What are the initiatives that can be taken to ensure that the military are able to obtain access to ECT technologies? This chapter has identified three strategies that the military could adopt to gain access to ECT which are the dual-use technology strategy, open-innovation, and offsets.

Moving forward

Dual-use technology strategy

Dual-use[18] defense technology is said to be a more effective solution to maintaining high-tech defense technology and economic competitiveness.[19] The application of dual-use technology is increasingly viewed as a positive transfer mechanism to solve issues in the defense sector, which is restrained by limited budgets, and to improve a nation's economic competitiveness through a more efficient allocation. There are different mechanisms in which technology can cross borders between civil and military applications. First, dual-use transfer can occur within the same or between different business units in an organization and second the technology may require adaptation in order to be transferred from the civil to the military domain and vice-versa.[20] By investing in dual-use technology, the military can exploit the efficiencies generated through the use of common production lines for commercial and military products, reap the reduced costs resulting from larger scale production runs, and leverage industry's willingness to invest in commercially viable technologies.

Government technology strategy to encourage "open innovation"

Government initiatives encourage open innovation as a way to reduce military research and development cost and tap into the commercial sector. Open innovation in the automotive sector has become increasingly popular due to the effects of the financial crisis and budgetary pressures faced by the defense sector due to defense inflation and spending cut backs in R&D; the military is constantly monitoring new commercial developments looking for military application. The spin-on of technology is of growing importance to the army for domestic and export. Due to the nature of the commercial automotive sector, many of the emerging technologies today originate in the civil automotive industry. The ministries of defense should be open to the idea of open innovation and embrace and collaborate with civil automotive partners to maximize the research and commercialization output and outcome. In the UK, for example, the Technology Security Strategy, Defence Business Strategy, and the UK Defence Innovation Strategy (2008) documents have all encouraged open innovation to reduce cost. The High Value Manufacturing Catapult project in the UK was also part of such initiatives to encourage high-value innovation in the automotive sector.[21]

The formation of National Automotive Innovation Centres (NAIC) is another way to encourage collaborative research and to get industry, government, and academia to work together on common technology projects. Further, such innovation centers can be used for combined defense–civil technology partnering for research and innovation into new technologies in the automotive sectors. The NAIC is there to encourage dual-use technology and issues such as cross-fertilization. WMG at the University of Warwick, for example, has been awarded a grant for the value of £92 million pounds by the UK government to set up the National Automotive Innovation Centre and further industry funding with involvement of major industries like Jaguar Land Rover.[22] Governments can also look at how such initiatives can help SMEs grow and technology can be shared by the SMEs that cannot afford the technology.

For developing countries, offsets and government procurement can be utilized as a source to create the platform for open innovation to obtain emerging technologies and leap frog development. Offsets can operate as the catalyst to create the absorptive capability for industries in the newer technological field. At the same time, offsets are also used to address issues of affordability for developing countries as they can request such newer technology as part of the defense equipment sale. Offsets are seen as a quicker mode of acquiring or getting access to cutting-edge defense automotive technology for developing countries to revolutionize their military capability and move towards greater modernization of military vehicles in line with current security challenges.

Offsets

Offsets are an economic development tool used to seek technology from foreign suppliers as part of an international sale of equipment and services. More than 70 countries around the world have some form of offsets policy. Most of these policies emphasize the importance of technology acquisition as part of the sale. The technology transfer can occur in various forms including design, manufacturing, licensing, sub-contracting, co-production, and knowledge transfer in the form of educational know-how. Most governments view offsets to have an impact on industrial and technological development of a nation although there are many counter-arguments to this claim.[23] In the case of ECT, many nations have now started using offsets to obtain ECT. This practice is seen as being cost effective as the newcomers to the technology business environment do not have to reinvent the wheel and are able to leap-frog in acquiring this new technology. There is an increasing pattern of developing countries in the Asia Pacific region using offsets to leverage ECT and this could be an increasingly popular trend for quick absorption of ECT in the future.

Conclusion

The interests in land-based technology innovation has taken prominence in recent years due to the new types of threats faced by the military operating in a land-based warzone. New and emerging technologies in the automotive sector originate from the civil and defense sphere though technology innovation is more rapid in the civil sector. The various implementation challenges have dampened the opportunity for full exploitation of these technologies by the armed forces. This chapter concludes that a government technology policy strategy focused on a "dual-use approach", "open innovation," and "offsets" could be the way forward to a more effective and efficient utilization and absorption of emerging technologies to enhance the land-based military capability. The discussion on utilizing all or any of the three strategies above is an option available besides the many other strategies. This is simply a future research into exploring the relevance and effectiveness of the dual-use strategy, open innovation, and offsets to leverage ECT for advancement of the defense automotive sector.

Notes

1. SIPRI yearbook defines armored vehicles as all vehicles with integral armor protection, including all types of tank, tank destroyer, armored car, armored personal carrier, armored support vehicle, and infantry fighting vehicle. Vehicles with very light armor protection (such as trucks with an integral but lightly armored cabin) are excluded. SIPRI Yearbook 2011, Armaments, Disarmament and International Security, Oxford University Press, 2011, p. 294.
2. J. Andrew (2013) " 'Policy Brief: Emerging Technology and Military Capability', Rajaratnam School of International Studies," Nanyang Technology University. Accessed through http://www.rsis.edu.sg/wp-content/uploads/2014/07/PB/31101-Emerging-Technologies-Military-Capability.pdf, November.
3. Accessed through http://clinton/nara.gov/white_House/EOP/OSTP/CT, dated 15 December 2014.
4. S.J. Richards (1994), *Rich Nation, Strong Army: National Security and Technological Transformation of Japan* (Ithaca, NY: Cornell University Press).

5. T. Shankar (2009), "Makeshift Bombs Spread Beyond Afghanistan, Iraq," *The New York Times*, October 28, http//:www.nytimes.com, accessed December 2012.
6. National Audit Office, Briefing for the House of Commons Environmental Audit Committee (2012), December (nao.org.uk, accessed January 2013).
7. UK Security Strategy (2010), "A Strong Britain in an Age of Uncertainty: The National Security Strategy," presented to Parliament by the Prime Minister by Command of Her Majesty, http://www.gov.uk, accessed on 25 February 2012.
8. National Security through Technology: Technology, Equipment and Support from UK Defence and Security. Presented to Parliament by the Secretary of State for Defence by Command of her Majesty, February 2012, http://www.gov.uk, accessed 18 June 2012.
9. Defence Technology Strategy for the demands of the 21st Century, MOD, Science Innovation Technology (2006), http://www.science.mod.uk, accessed 18 June 2012.
10. Disruptive Innovation is an innovation that helps create a new market and value network and eventually disrupts an existing market and value network, displacing an earlier technology.
11. For open innovation see H. Chesborough, "Open Innovation: A new paradigm for understanding industrial innovation," in H. Chesborough, W. Vanhaverbeck and J. West (Eds) (2006), *Open Innovation: Researching a New Paradigm* (Oxford: Oxford University Press).
12. I. Gavin (2008), "Accelerating Defence Acquisition: What Defence can learn from the World of Motorsport", *RUSI*, June, pp. 80–83.
13. Interview with MIRA official. MIRA is a UK based automotive research center that specializes in automotive technology research, development and commercialization.
14. UK MOD Sustainability Development Report (2013), http://www.gov.uk, accessed on 15 May 2014.
15. See *The Economist*, "A Third World Dimension, A New Manufacturing Technique Could Help the Poor Countries as Well as Rich Ones," dated 3 November 2012; *The Economist*, "Print me a jet engine", dated 22 November 2012.
16. Technology Strategy Board, https://www.innovateuk.org/transport, dated 11 May 2014.
17. Batteries and battery packs (high performance battery modules, packs and cells, low-cost battery management system, recycling of battery products, and reducing overall carbon footprint, battery test facility; drive motors; research into lightweight structures to reduce overall weight of vehicle and improve fuel economy, while maintaining desired levels of vehicle performance targets; lightweight defense vehicle and lightweight structures; defense vehicle braking; energy storage and management; defense supply chain; ED and blast protection; batteries and battery packs; power electronic; high voltage electrical distribution; auxiliary power units; vehicle supervisory control; electric vehicles (energy storage and e-drives); carbon reduction.
18. The term dual use refers to product, knowledge and skills, and activities in developing technology. The difference in civil or military application will

be determined by the difference in government infrastructure and regulations, goals (national security versus commerce), market structure, standard specifications ("milspecs"), sensitivity to costs, different product cycles, and industry and technological cultures.

19. H.T. Kulve and W.A. Smit (2003), "Military Cooperation Strategies in developing new technologies", *Research Policy*, vol. 32, pp. 955–970.

20. J. Molas-Gallart (1997), "Which Way to Go? Defence Technology and the Diversity of Dual-use Technology Transfer", *Research Policy*, vol. 26, pp. 367–385.

21. The High Value Manufacturing Catapult (HVM) project is a UK-wide initiative to bring together the best scientist and engineers based in the UK to help accelerate new concepts to commercial reality. The HVM catapult consists of seven consortia of world-class research centers with over 160 industries and other partners. The funding for this initiative involved at least £75million per annum with a 50–50 split of public and private partnership funding.

22. Interview with a WMG, University of Warwick official and also refer to http://www2.warwick.ac.uk/fac/sci/wmg/about, dated 16 May 2014.

23. B. Kogila and M. Ron (2009), "Malaysia's Defence Industry and the Role of Offsets," *Journal of Defence Peace and Economics*, vol. 20, no. 4, August, pp. 342–355.

7
Diesel-Electric Submarine Modernization in Asia: The Role of Air-Independent Propulsion Systems

Michael Raska

An important aspect of the regional "arms competition" in East Asia is the gradual introduction of new classes of conventionally-powered diesel-electric submarines, which are increasingly becoming "platforms of choice" – as force-multipliers in diverse missions as well as against superior forces. Coupled with submarine-launched anti-ship and land-attack cruise missiles, advanced intelligence, surveillance, and reconnaissance (ISR) sensors, anti-submarine sensors and weapons, as well as new propulsion systems – such as air-independent propulsion (AIP) – these new classes of submarines have a greater capacity to remain undetected (stealth), with improved target-identification-and-attack cycle and ultimately increased mission flexibility, mobility, endurance, reach, and lethality.[1] In particular, conventional submarine modernization and expansion has been profound in Northeast Asia, driven by multiple factors, including re-capitalization, replacement, reactive acquisitions, territorial and maritime boundary disputes, and great power aspirations.[2] Indeed, there are at least three underlying drivers that support and accelerate submarine procurement and diffusion among Asian navies: (1) the persisting geopolitical insecurity, regional rivalry, and uncertainty stemming from the nexus of unresolved historical legacies to emergence of complex types of conflicts that broaden national defense requirements and operational needs; (2) increased regional economic growth rates that increase the capacity to purchase cutting-edge weapons systems and accelerate force modernization programs; and (3) the rapid technological change embedded in the globalization, consolidation, and competition of global arms markets and defense industries that have to diversify their commercial interests through export-oriented strategies and innovation – particularly aiming at Asia-Pacific markets.

However, the primary strategic driver for submarine procurement in East Asia is arguably China's ongoing qualitative military modernization. The economic, political, and military rise of China, reflected in three decades of relentless Chinese economic growth, has propelled progressive modernization of the Chinese military with major improvements in virtually every capability domain: land, air, naval, missile, space, cyber, and electronic warfare.[3] The cumulative effects of these developments are substantial as China's catalogues of air, land, and naval platforms are gradually catching-up in terms of both qualitative sophistication and operational effectiveness.[4] In March 2014, China announced a $131 billion defense budget, up 12.2 percent from the 2013 budget of $119 billion – marking 17 straight years of near-double digit increases in defense spending.[5] As China expands its national interests in the broader context of "new historic missions," it seeks to regain a great power status and reassert its geopolitical role in the region. China's People's Liberation Army (PLA) is gradually pursuing capabilities for force projection into China's "near seas" or an area defined by the "first island chain" consisting of the Kuril Islands, Japan, Taiwan, and the South China Sea. In this context, China's PLA Navy (PLAN) is gradually transforming toward a "regional [blue water] defensive and offensive type navy with extended so-called anti-access/area-denial (A2/AD) capabilities, limited expeditionary capabilities, and corresponding defensive and offensive air power."[6] China's calls its comprehensive A2/AD strategy a "counter-intervention," which is interpreted in the US strategic thought as denying US forces the freedom of action in China's "near seas" by restricting deployments of US forces into theatre (anti-access) and denying the freedom of movement of US forces already there (area denial).[7] In the long term, China envisions capabilities that would extend its strategic reach into the "second island chain," which includes US bases on Guam. Chinese air power doctrine, for example, envisions the PLA Air Force by 2030 conducting independent air campaigns within 3,000 km radius of China's periphery – shifting its primary missions from traditional land-based air defense, interdiction, and close air support operations, toward deterrence and strategic strike at sea.[8]

In this context, China is focusing on the procurement and modernization of its submarine fleet. The PLA Navy currently operates as many as 45 submarines structured in six different classes: two classes of indigenously designed diesel submarines, including the *Song*-class

(*Type 039*) and the *Yuan*-class (*Type 041*), and four nuclear classes that include the *Shang*-class (*Type 093*), *Jin*-class (*Type 094*) nuclear-powered ballistic missile submarines (SSBN) and the follow-on *Type 095* nuclear-powered attack submarine (SSN) and *Tang*-class (*Type 096*) SSBN.[9] Since 2004, China is believed to have launched 12 *Type 041* conventional submarines, which have been progressively modified to carry advanced high-frequency sonar technologies as well as more weapons systems.[10] The PLA Navy may procure up to 20 additional *Yuan*-class submarines,[11] based on German diesel engine purchases, and technologies imported from Russian boats. Since the mid-1990s, China has procured as many as 12 *Kilo*-class submarines from Russia, and is reportedly negotiating the purchase of at least four fourth-generation *Amur* (export version of the *Lada*-class) and possibly a fifth-generation *Kalina*-class, both featuring advanced AIP systems.[12]

Taking into account the strategic ramifications of China's ongoing naval modernization and power projection, Japan and South Korea are also modernizing and expanding their submarine forces. In July 2014, South Korea launched a fifth 1,800-ton *Son Won-ill* class ROKS *Yun Bong-gil* (based on German *Type-214* submarine) – a diesel-electric, AIP submarine equipped with *Haeseong-3* missile, and advanced combat management systems. With the new boat, South Korea currently operates 14 submarines: nine *Type 209 Chang Bogo* and five *Son Won-ill* class submarines.[13] The country also entered a design phase for a new, indigenous 3,000-ton submarine, equipped with a vertical launch missile capability. The South Korean navy plans to acquire nine of these submarines between 2020 and 2030.[14] Meanwhile, in October 2013, the Japan Marine Self Defense Force (MSDF) launched its newest submarine: the *Kokuryu* – the sixth in the latest *Soryu*-class. With its range, endurance, sensors, weapons load, and other systems, including the Stirling AIP system and Harpoon anti-ship missiles, the *Soryu*-class is regarded as the most advanced in Japan's current conventional submarine fleet of 16 submarines.

While the relatively high acquisition costs and maintenance requirements have precluded a quantitative diffusion of submarines in operational fleets in Southeast Asia, the qualitative dimension in the recent introduction of more capable coastal diesel-powered submarines provides new capability. In 2013 and 2014, Vietnam received two of six *Kilo*-class (Project 636) diesel-electric submarines from Russia, designed for diverse reconnaissance and patrol, anti-submarine and anti-ship missions. By 2018, with all six *Kilos* operational and equipped with *Klub-S* (3M-54) anti-ship cruise missiles, Vietnam's subsurface fleet could potentially represent the largest undersea force in the region.[15] Indonesia, Malaysia, and Singapore are also planning to expand or upgrade their submarine fleets. From 2007 to 2009, Malaysia took a formal delivery of two French-built *Scorpene*-class submarines, equipped with underwater-launched *SM-39 Exocet* anti-ship missiles. Both submarines are based at the Kota Kinabalu Naval Base in Sabah, East Malaysia, indicating their primary mission to protect Malaysia's sovereignty in the contested waters of the South China Sea.[16] Meanwhile, Indonesia has ambitious plans to expand its submarine fleet to 12 by 2024, a key element in the "Minimum Essential Force" (MEF) and declared goal of developing a "green-water" navy.[17] In 2012, the Indonesian Navy (Tentera Nasional Indonesia-Angkatan Laut or TNI-AL) announced a US$1.1 billion contract for three *Type-209/1400* diesel-electric submarines, constructed by South Korea's Daewoo Shipbuilding and Marine Engineering. The new submarines will provide incremental technological and capability upgrades over the two existing German-built *U-209/1300* submarines, in service since 1981.[18] And in November 2013, Singapore announced a contract with German shipbuilder ThyssenKrupp to acquire two advanced *Type-218SG* submarines that will augment existing *Archer*-class boats and replace ageing ex-Swedish *Challenger*-class by 2020. *Type-218SG*, designed for littoral, shallow-sea operations, is a customized design based on a new German-*Type-216* "concept submarine" and will be fitted with AIP from the baseline.[19] Last but not least, Thailand, Philippines, and Myanmar have also evinced interest in acquiring undersea warfighting capabilities; however, their actual implementation is limited by ongoing budgetary constraints.[20]

Why submarines?

These trends reflect the East Asia's changing strategic realities, shaped by contending trajectories of unresolved historical legacies and emerging security challenges. East Asia's strategic template is shifting toward a mix of asymmetric anti-access/area-denial (A2/AD) threats, low–high intensity conventional conflicts in traditional flashpoints such as the Taiwan Strait or the Korean Peninsula, and a range of non-traditional security challenges such as energy security, cybersecurity, and intraregional competition in territorial disputes in East China and the South China Seas. East Asia's "hybrid" security challenges coupled with the diffusion of advanced defense technologies in nearly every battlefield domain is also widening operational requirements and compels states to adopt and adapt military capabilities and innovations of their existing or potential rivals. The United States, Japan, South Korea, and to a lesser degree Taiwan are acquiring new power projection capabilities – from reduced-signature fifth-generation air platforms, standoff precision weapons, ballistic and cruise missiles, early warning, ISR assets to naval assets, including maritime patrol, anti-submarine warfare (ASW), and submarines. At the same time, they are demonstrating the political willingness to use these assets for different strategic reasons. Japan seeks to overcome the limitations posed by its pacifist postwar constitution and the Yoshida Doctrine to exercise greater strategic adaptability and operational flexibility in responding to regional contingencies. Tokyo is rethinking its national defense posture and its overall security role in the region. In November 2013, Japan's cabinet under Prime Minister Shinzo Abe launched the country's first National Security Council, followed by the approval of the first National Security Strategy, and increased efforts to reinterpret Japan's pacifist constitution.[21] Meanwhile, South Korea's ongoing defense reforms and acquisition programs have aimed not only at strengthening capabilities vis-à-vis North Korean asymmetric threats, but also developing joint air and naval capabilities that would complement long-term US strategic interests in East Asia.[22] The United States, with its policy of strategic rebalancing toward the Asia-Pacific, seeks to remain a "Pacific Power" through economic, diplomatic, cultural, and military presence and influence.[23]

These trends are also altering the strategic context in contested areas in Southeast Asia, particularly in the South China Sea. Smaller and medium-sized states in Southeast Asia are gradually modernizing their naval and air forces to keep their vital sea lanes open, conduct intelligence missions, and perhaps most importantly, provide credible asymmetric "force multipliers" in deterring and potentially disrupting China's naval forces from seizing disputed islands in the South China Sea. While there are different political, strategic, and technological drivers shaping regional military modernization trajectories, including long-standing intraregional rivalries and competition over borders, resources, and history, most Southeast Asian countries share concerns about China's military capabilities and future aspirations in the region. As a result of China's "coercive diplomacy" since mid-2010, Southeast Asian countries are responding by revamping their force-modernization priorities, alliances, and overall strategic choices. The result is not a regional "arms race" per se (i.e. action-reaction cycle of arms acquisitions based on mutually adversarial relationships, explicit tit-for-tat arms acquisitions, the intention of seeking dominance over one's rivals through arming and intimidation), but rather a gradual "arms competition" or "arms dynamic" characterized by a mix of cooperative and competitive pressures, continued purchases of advanced weapon platforms, including the introduction of new types of arms and, therefore, unprecedented military capabilities.[24] In particular, for smaller, defensively-oriented navies in East and Southeast Asia, submarine stealth attributes enable "sea-denial" capabilities – "preventing an opponent from using the sea, rather than providing a degree of sea control that would enable them to use the sea themselves for power projection or trade projection."[25] In other words, the objective of sea-denial is not to use the sea oneself, but to prevent the enemy from doing so.[26] Stealth attributes of submarines provide strategic advantages over surface ships as the means for sea-based surveillance, particularly in the shallow coastal and littoral waters in the South China Sea and Java Sea. Even if submarines are detected, it may not be always possible to identify their type and nationality, which may consequently complicate the response options available to opponents.[27]

Emerging critical technologies: AIP systems

One of the key emerging technologies incorporated in the latest designs and classes of conventional submarines in East Asia are select AIP systems. A conventional submarine, powered by diesel engines, must use oxygen sourced from the atmosphere when the engines are running. These submarines must therefore travel on the surface or at a shallow depth with a snorkel above the surface to take in oxygen and expel combustion fumes. Without snorting, a submarine can stay beneath the surface for an average of only about 100 hours, operating at 4 knots on battery power. With an AIP system on board, however, a submarine can extend the time between snorkel operations to recharge its batteries up to two weeks or more, depending on its speed. While snorkels have improved with radar absorbent materials and shapes to minimize their bow wave and wake signatures, they emanate a detectable signature – and are increasingly vulnerable to advanced counter-detection systems.[28] In essence, AIP systems provide a "closed-cycle" operation through a low-power electrical source that *supplements* the battery without the requirement of an external air source. AIP systems can therefore vastly extend the range of possible mission scenarios of submarines of more than two weeks, while decreasing the risks of being detected.

AIP systems, however, are not new. The intellectual roots of AIP technologies emerged in the latter years of World War II, when the

German *Kriegsmarine* faced a threat of well-armed Allied maritime patrol aircraft, capable of detecting German diesel-electric U-boats on the surface. As a countermeasure, German engineers developed the snorkel, which permitted the U-boats to draw air to recharge the batteries while partially submerged. In 1944, they also experimented with new submarine propulsion devices using a concentrated hydrogen peroxide (High Test Peroxide HTP), designed primarily to increase underwater speed.[29] Their ambitious plans were thwarted by the unsuccessful course of the war, industrial limitations, and various technical and safety problems. One of their designs – *Type XVIII*, however, was modified into *Type XXI* "electro-boat," which significantly influenced the development of conventionally-powered Soviet, British, and US submarines during the early years of the Cold War.[30] While major naval powers subsequently shifted to nuclear propulsion, smaller navies – particularly in Europe (Germany, Sweden, Spain, Italy, and France) have continued to develop and rely on conventional diesel-electric submarine fleets, given their lower cost and operational relevance for coastal defense. Traditionally, however, these submarines were highly vulnerable to various types of acoustic, visual, thermal, and air sensors, particularly when running on engines. On the other hand, when running on batteries, these submarines became very quiet and difficult to detect, yet their battery capacity, discharge rate, and indiscretion rate (the ratio of diesel running time to total running time) substantially limited their underwater endurance.[31] Generally, in a battery-powered mode, faster speeds (over 15 knots) would require higher power-consumption and accelerate power-drain, which limited the operational range. Subsequently, this also limited the options for installed onboard systems, weapons load, and overall performance. To overcome these baseline limitations, close the endurance gap between nuclear and conventional submarines, and mitigate increasing risks of detection provided by advanced ASW technologies (from modern electro-optical systems and surface radars to magnetic sensors, active and passive sonars, and airborne surveillance radars), naval innovation in propulsion technologies over the past two decades has shifted toward AIP systems. Advanced AIP technologies promise significant operational advantages in the prolonged submerged endurance capability that enables submarines to operate at low speeds for two to three weeks without the need to snort; lower acoustic, bow wave, and wake signature in the water, and the resulting tactical flexibility coupled with lowered probability of detection and expanded operational capabilities.

Seen from this perspective, there are five key questions that define the main analysis of this chapter: (1) to what degree can AIP systems represent a critical type of innovative technology or emerging capability in terms of creating comparative advantages for a nation's military over a potential rival in East Asia? (2) What are the key parameters and indicators for comparison, evaluation, and integration of AIP systems? (3) What factors are likely to drive technological advances in AIP systems in East Asia? (4) Does possessing a strong S&T base or having access to AIP technologies naturally lead to possessing a military advantage? (5) How might the unequal distribution of AIP systems affect military capabilities in the Asia-Pacific? Are there "offsetting" alternatives that a country may pursue in order to asymmetrically compete with a more technologically advanced rival? While providing a limited set of answers, this chapter projects that the *net effectiveness*, both at the strategic and technological levels, of select AIP technologies can be assessed at two levels: (1) *Systemic* – based on a qualitative portfolio of all available submarine assets – platforms, systems, and technologies coupled with innovative operational concepts designed to perform new missions or significantly improve old systems or processes to achieve select maritime objectives; (2) *Functional* – based on a portfolio of "capability domains" or functional tasks in a particular set of operational missions or undersea force deployments in which an AIP system would enable regional navies to operate at higher levels of operational complexity.[32] These capability domains include factors such as submerged endurance, suitability of the AIP systems to operating environments, acoustic signature in specific operating regimes and at varying speeds and depths, vulnerabilities and failure modes of the AIP systems, and procurement and life-cycle costs in relation to the total cost of the submarine.[33] In theory, there are four primary AIP designs currently available: (1) closed-cycle diesel engines; (2) closed-cycle steam turbines; (3) Stirling-cycle heat engines with external combustion, and (4) hydrogen-oxygen fuel cells.[34] Each provides a different solution with particular technological advantages as well as operational limitations in relation to overall performance, safety, and costs.

Closed-cycle diesel engines (CCD AIP)

In theory, one of the most affordable systems is the CCD AIP design, which is based on a modified standard diesel engine that can operate in its conventional mode on the surface or when snorkeling, and as a closed gas cycle underwater. The CCD generates artificial atmosphere synthesized from stored liquid oxygen, an inert gas (argon), and recycled exhaust products. The engine exhaust (i.e., carbon dioxide, nitrogen, and water vapor) is cooled, scrubbed, and separated into its constituents, with the argon recycled back. The remaining exhaust gas is mixed with seawater and discharged overboard. Early versions of the CCD were adopted by the Soviet Navy; however, their limited endurance, malfunctions, and safety problems in storing the liquid oxygen forced them to abandon the program in the 1970s.[35] From 1989 to 1993, German Thyssen Nordseewerke GmbH developed its CCD AIP system and successfully tested it on board the experimental *Ex-U1* (former *Klasse 205* U-boat). Similarly, in the 1990s RDM submarines (Netherlands) installed a *CCD Spectre* system on board the decommissioned Dutch submarine *Zeehond*, and currently offers the system for the 1,800 ton hybrid-powered *Moray H* submarine. While these systems have been offered to the Argentinean Navy, the Royal Netherlands Navy, the German Navy, the South Korean Navy, and the Egyptian Navy – none of them have been procured. Notwithstanding their relative low-cost, minimal changes to interoperability or support infrastructure, the key limitations of CCD are four-fold: (1) the CCD requires a "plug-in" section to contain the argon and oxygen tanks and other components, which may adversely influence the boat's maneuverability; (2) the CCD generates similar noise levels and heat as regular diesel-electric motors, which makes them vulnerable to detection by sonar and infrared means; (3) the CCD ejects dissolved exhaust, which could be detected by chemical sensors; and (4) the CCD overall performance is similar to conventional diesel-electric engines.

Closed-cycle steam turbines

Developed by the French DCNS International naval shipbuilding company, the only AIP steam-turbine system currently available is the "MESMA" (*Module d' Énergie Sous-Marine Autonome*) module. Its principle mechanism is a steam-cycle used in nuclear plants (and employed by French submarines and aircraft carriers) and adapted to non-nuclear steam generation – a conventional Rankine-cycle turbo-electric generator, powered by the steam generated from the combustion of ethanol and stored liquid oxygen. DCNS claims the MESMA increases submarine underwater endurance by a factor of 3–5 at a speed of 4 knots – with extended submerged endurance by 21 days. The company also notes the low life-cycle cost and maintenance requirements similar to any steam engine. DCNS offers the MESMA option for its *Agosta 90B* and *Scorpene* classes of submarines – requiring the insertion of an 8.5 meter 305-ton hull section, which can be used for existing submarines.[36] Outside of

France, MESMAs are operational only on two of Pakistan's Navy *Agosta 90-B* class submarines; following the 1994 contract, the first MESMA was retrofitted in 2008 on board the *Hamza* submarine. The second system was delivered in 2012, and the third in 2013.

Notwithstanding its relatively higher power output than other AIP systems, the MESMA has inherent operational constraints. These include the requirement to store and handle LOX, which increases potential safety hazards. The MESMA also contains moving parts, which may radiate detectable noise-levels. Moreover, the burning process generates exhaust carbon dioxide that must be expelled behind the submarine at any depth, making it vulnerable to advanced airborne and ASW ship sensors. At the same time, its fuel cell efficiency might be the lowest (est. 25 percent) as its rate of oxygen consumption is higher.[37] The maintenance costs and crew training requirements for the MESMA steam turbine system are also significant compared to other AIP systems.

Stirling-cycle heat engines

In 1988, Swedish Kockums Naval Systems installed the world's first Stirling engine prototype AIP system in the Royal Swedish Navy submarine *HMS Nacken*. Stirling engines (invention dating back to 1816) burn pure oxygen and diesel fuel in a pressurized combustion chamber. The combustion pressure is higher than the surrounding seawater pressure, which allows the exhaust products, dissolved in seawater, to be vented directly to sea without using a compressor. The oxygen is stored in liquid form (LOX) in separate cryogenic tanks. The Stirling engine is linked to a generator that feeds into the submarines' primary electrical system.[38] According to Saab Kockums, Stirling AIP extends submerged endurance from a few days up to 18 days, depending on the submarine's speed and levels of stored LOX.[39] Currently, the Royal Swedish Navy operates five Kockums-built submarines equipped with Stirling AIPs. These include the *Gotland class* (*HMS Gotland*, *HMS Uppland*, and *HMS Halland*) that feature two 75 kW Stirling engines for propulsion or charging batteries, and two *Södermanland class* submarines (*HMS Södermanland* and *HMS Östergötland*) that have been upgraded with the Stirling AIP. Sweden's next generation submarine, Kockums *A26* – currently in development, will incorporate the latest and most modern and refined Stirling AIP technology – the Mk 5 version. Outside Sweden, Stirling AIP is used in two of Singapore's Navy *Archer class* submarines (modernized ex-Swedish Navy *Västergötland class* submarines). In 2005, Japan's Kawasaki Heavy Industries signed a contract with Kockums for a licensed production of Mk III Stirling AIP system for Japan's *Soryu class* submarines, which utilize four 75 kW Kawasaki Kockums V4-275R Stirling engines. The Chinese PLA Navy's *Type 041 Yuan* and *Type 043 Qing* class submarines are also reportedly using the Stirling technology.

As with other AIP systems, Stirling engines have both operational advantages as well as limitations. While these systems are relatively simple, less prone to safety hazards, and featuring advanced noise-reduction technologies such as double-elastic mounting with a sound proof hood to minimize noise levels, their supporting auxiliary equipment contains moving parts that contribute to noise and limit the submarine's stealth capabilities. Stirling engines operate at a pressure of 20 bars, which limits the submarine's depth capability to 200 meters, unless a power consuming and potentially noisy exhaust gas pressure intensifier is used. Ultimately, as any other AIP system burning diesel and LOX, the LOX supply will remain the principal constraint to achievable endurance.[40]

Fuel cell systems – PEM AIP

Fuel-cell-based AIP systems utilize a hydrogen–oxygen fuel cell to generate electrical current that powers the submarine systems without noise or combustion. They operate by recharging the submarines' batteries or in the case of Polymer Electrolyte Membrane (PEM) technology directly feed the electrical motor, leaving only distilled water as a waste product.[41] Fuel cells have been already applied in space vehicles; their major advantage is that the power capability of the fuel cell stack can be varied independently of the energy storage capability.[42] Moreover, there is no exhaust gas production – the only "exhaust" product is pure water – which is retained on board for weight compensation purposes. Other key advantages are absence of moving parts and absolute silence in the energy generation process, high efficiency of up to 70 percent of energy, depth independence, low operation temperature (80C), short start-up time, modular design, and its flexibility and automation in operational and control features.[43] Notwithstanding these operational advantages, fuel cells AIP systems also have inherent weaknesses: high procurement costs, challenges in retrofitting the PEM system into existing submarines; and safety hazards in the physical and chemical storage for the onboard oxygen and hydrogen, which could be fraught with fire or explosion.

Development of the first generation of fuel cell plants based on alkaline fuel cells (AFC) and later polymer electrolyte membrane (PEM) fuel cells for submarines began in the early 1980s in Germany by a consortium of Howaldtswerke-Deutsche Werft AG (HDW), Ingenieurkontor Luebeck (IKL), Siemens, and Ferrostaal. In the 1990s, the German Navy decided to adopt the system for the *Type 212A* submarines with nine 30-50kW fuel cell units, followed by the export-designed *Type 214*.[44] Since 2002, fuel cell AIP systems have been successfully integrated into many German-made export submarines. The Republic of Korea Navy has ordered nine *Type 214* submarines, designated as *Son Won-Il class*, built in Korea by Hyundai Heavy Industries and Daewoo Shipbuilding & Marine Engineering; three first-batch models had entered service since 2007, and six second-batch models will enter service from 2012. Other countries procuring the 214 variants include Greece, Portugal, Pakistan, and Turkey. Meanwhile, Israel has procured five *Type 800 Dolphin* and *Dolphin Class II* submarines from Germany, which employ two state-of-the-art 120 kW fuel cell AIP units.[45] As for Singapore's *Type 218SG* submarines, currently under construction in Germany, it is unknown whether these will employ fuel cell PEM or Stirling engines as in the existing two Archer class submarines.[46] An alternative to German fuel cell technology is the *S-80 Class* submarine, currently under construction by the Spanish company Navantia. The Spanish AIP system is based on a fuel cell using liquid oxygen and hydrogen extracted from bio-ethanol, which allows the boat to remain submerged longer than with batteries. Navantia claims it has the highest AIP range in the market as the initial two-week submersion endurance in AIP could be extended.[47]

Strategic and operational ramifications

With emerging AIP technologies, the mission templates of submarine activities in East Asia has significantly widened: from ASW to force protection such as close submarine escort missions, ISR, support of special forces, and other complementary deterrence and defensive tasks supporting territorial defense. China's ongoing military modernization coupled with contending international relations in the region will further drive submarine procurement over the next decade. In 2011, the total submarine market in Asia-Pacific was estimated at US$4.4 billion, and for the next decade, submarine expenditures are projected to rise to US$46 billion.[48] These trends will force East Asian submarine navies to become increasingly flexible, and capable of varying mission profiles: from countering traditional coastal defense missions to protecting sea lanes and communication lines. Simultaneously, submarines will be an increasingly valuable strategic resource for both electronic and signal intelligence. In order to enhance the varying operational capabilities, increase submerged endurance and stealth, installing viable AIP systems *or* their alternative technologies such as ion-lithium battery propulsion systems will become a necessity. Another key factor that will likely propel AIP diffusion in Asia are advances/digitization of ASW sensor technologies concomitant with Synthetic Aperture Radar (SAR) techniques detecting wakes of submerged submarines, LIDAR (laser radar) technologies, electro-optical sensors, and ELS/ESM (Emitter Locating System/Electronic Support Measures.[49] Developments in sonar, radar, and other areas will improve the effectiveness of ASW sensor suites, which may also provide viable countermeasures to the various AIP systems.

At the same time, however, the greatest constraint to AIP technologies will lie in their own technological limitations, vulnerabilities, and risks, particularly in submerged operations – from their specific acoustic signatures produced by AIP systems in specific operating regimes, to technical vulnerabilities in storing oxidizer/fuel, as well as their maintenance regime. In this context, any critical failure of an AIP system during a combat mission or contested areas will inherently mitigate survivability factors as well as tactical options.[50] Therefore, the effectiveness of AIP propulsion systems will also depend on the wide array of other key systems that must be integrated to ensure optimal submarine functions: power systems, sensors systems, safety systems, navigation systems, command, control, and communication systems, weapons systems, and climate control systems. Inherently, the new classes of conventional submarine designs are complex in terms of defense management, logistics, maintenance of equipment and infrastructure. At the core of each new submarine acquisition is not only its strategic utility, but also attaining efficiency gains, cost savings, experience and expertise in select areas of defense management, financial planning, training doctrine, and supporting organizations. Consequently, operationally-ready submarines will require a long-term organizational investment, highly-skilled workforce, and relevant doctrine – a modern navy that can produce commanding officers with up to ten years of experience, senior sailors, operators or maintainers capable of managing the submarines' weapons, propulsion, and communications systems. These include imperatives to attain a qualitative technological expertise, organizational learning, and operational proficiency by highly skilled and experienced operators and technicians. As with other emerging technologies, AIP-related technological innovation and breakthroughs in themselves may not guarantee operational success – strategy, operational concepts, tactical development, leadership, and training will continue to play as important a role as technological evolution and advancement.

Notes

1. Till, Geoffrey. 2013. *Seapower: A Guide for the Twenty-First Century (Third Edition)*. London: Routledge. p. 125.
2. Storey, Ian. 2014. "Naval Modernization in China, Japan, and South Korea: Contrasts and Comparisons." In *Naval Modernization in South-East Asia: Nature, Causes, and Consequences*, by Geoffrey Till and Jane Chan (eds), 104–109. New York: Routledge.
3. Tellis, Ashley. 2012. "Uphill Challenges: China's Military Modernization and Asian Security." In *Strategic Asia 2012–13: China's Military Challenge*, by Ashley Tellis and Travis Tanner (eds), 3–24. Washington D.C.: The National Bureau of Asian Research.
4. Erickson, Andrew. 2012. "China's Modernization of its Naval and Air Power Capabilities." In *Strategic Asia 2012–13: China's Military Challenge*, by Ashley Tellis and Travis Tanner (eds), 60–125. Washington D.C.: The National Bureau of Asian Research.
5. Minnick, Wendell. 2014. "China Mixing Military Modernization with Tailored Coercion." *Defense News*, 14 April, p. 11.
6. Li, Nan. 2009. "The Evolution of China's Naval Strategy and Capabilities: From 'Near Coast' and 'Near Seas' to 'Far Seas'". *Asian Security* 5(2): 168.
7. Schreer, Benjamin. 2013. *Planning the Unthinkable War: "AirSea Battle" and its Implications for Australia*. Barton: Australian Strategic Policy Institute. p. 8.
8. Mastro, Oriana Skylar and Stokes, Mark. 2011. "Air Power Trends in Northeast Asia: Implications for Japan and the U.S.-Japan Alliance." *Project2049 Institute Occasional Papers*. p. 4.
9. Erickson, "China's Modernization of its Naval and Air Power Capabilities," pp. 67–68.
10. Fisher, Richard, Hardy, James, and Rahma, Ridzwan. 2014. "China's New Yuan-class Sub Seen Preparing for Sea Trials." *IHS Jane's Defence Weekly*, 9 April.
11. U.S. Department of Defense. 2013. *Military and Security Developments Involving the People's Republic of China 2013*. Annual Report Prepared for Congress, Washington, D.C. p. 7.
12. Keck, Zachary. 2014. "Russia May Sell China New Advanced Submarines." *The Diplomat*, 28 March. Available at: http://thediplomat.com/2014/03/russia-may-sell-china-new-advanced-submarines/ [Accessed Date: 14 October 2015].
13. Schreer, Benjamin. 2013. "South Korea's Developing Blue Water Navy." *ASPI Strategist*, 6 September. Available at: http://www.aspistrategist.org.au/south-koreas-developing-blue-water-navy/ [Accessed Date: 14 October 2015].
14. Yonhap News Agency. 2013. "Navy to Build Nine 3,000-ton Subs by 2030: Source." *Yonhap*, 4 August. Available at: http://english.yonhapnews.co.kr/national/2013/08/04/30/0301000000AEN20130804001000315F.html [Accessed Date: 17 October 2015].

15. Koh, Swee Lean Collin. 2012. "Vietnam's New Kilo-class Submarines: Game-changer in Regional Naval Balance?" *RSIS Commentaries* (162): 1–2.
16. Beng, Aaron. 2014. "Submarine Procurement in Southeast Asia: Potential for Conflict and Prospects for Cooperation." *Pointer – Journal of Singapore Armed Forces* 40(1): 55–66.
17. Schreer, Benjamin. 2013. *Moving beyond Ambitions? Indonesia's Military Modernisation*. Barton: Australian Strategic Policy Institute. p. 19.
18. Koh, Swee Lean Collin. 2012. "Indonesia's New Submarines: Impact on Regional Naval Balance." *RSIS Commentaries* (21): 1–2.
19. Eshel, Tamir. 2013. "Singapore's Type-218SG: Forerunner of a new Submarine Class?" *Defense Update*, 5 December. Available at: http://defense-update.com/20131205_singapore_type_218sg-submarine.html#.ViQ2rvkrKHs [Accessed Date: 15 October 2015].
20. Supriyanto, Ristian. 2011. "Southeast Asia's Underwater Bazaar." *The Jakarta Post*, 26 July. Available at: http://www.thejakartapost.com/news/2011/07/26/southeast-asia%E2%80%99s-underwater-bazaar.html [Accessed Date: 14 October 2015].
21. Hardy, James. 2014. "Strong Constitution: Japan Looks to Reset Its Policy on Self-Defence." *IHS Jane's Intelligence Review*, 14 May.
22. Lee, Chung Min. 2003. "East Asia's Awakening from Strategic Hibernation and the Role of Air Power." *Korean Journal of Defense Analysis* 15(1): 219–274.
23. Ratner, Ely. 2013. "Rebalancing to Asia with an Insecure China." *The Washington Quarterly* 36(2): 21–38.
24. Bitzinger, Richard. 2010. "A New Arms Race? Explaining Recent Southeast Asian Military Acquisitions." *Contemporary Southeast Asia* 32(1): 50–69.
25. McCaffrie, Jack. 2014. "Submarines for South-east Asia: A Major Step?" In *Naval Modernization in South-East Asia: Nature, Causes, and Consequences*, by Geoffrey Till and Jane Chan (eds), 29–52. New York: Routledge.
26. Till, *Seapower: A Guide for the Twenty-First Century (Third Edition)*. p. 152.
27. McCaffrie, "Submarines for South-east Asia: A Major Step?" p. 40.
28. Kopp, Carlo. 2010. "Air Independent Propulsion – Now a Necessity." *Defence Today* 8(5): 10–12.
29. The Kriegsmarine built three Type XVIIB costal submarines (300-ton displacement), and planned to build larger Type XXVI (800-ton) Type XVIII (1,600-ton) U-boats.
30. Whitman, Edward. 2001. "AIP Technology Creates a New Undersea Threat." *Undersea Warfare* 4(1): 7–13.
31. Stefanick, Tom. 1987. *Strategic Antisubmarine Warfare and Naval Strategy*. Lexington: D.C. Heath. pp. 144–145.
32. Paul K. Davis, Russell D. Shaver, and Justin Beck, 2008. *Portfolio-Analysis Methods for Assessing Capability Options*. Santa Monica: RAND; Davis, Paul. 2002. *Analytic Architecture for Capabilities-based Planning, Mission-System Analysis, and Transformation*. Santa Monica: RAND.
33. Kopp, Carlo. 2010. "Air Independent Propulsion – Now a Necessity." *Defence Today* 8(5): 10–12.
34. Rex, Patrick. 2011. "Sea 1000: Conventional Air Independent Propulsion." *Asia-Pacific Defence Reporter* 37(10): 40–44.
35. Walsh, Don. 2000. "Air-Independent Propulsion: An Idea Whose Time Has Come?" *Seapower*, Navy League of the U.S. p. 36.

36. DCNS. 2010. "MESMA: AIP Module for Conventional Submarines." Available at: http://www.gican.asso.fr/content/mesma [Accessed Date: 17 October 2015].
37. Whitman, Edward. 2001. "AIP Technology Creates a New Undersea Threat." *Undersea Warfare* 4(1): 7–13.
38. Coates, Peter. 2014. "Air Independent Propulsion Technologies and Selection." *Submarine Matters*. Available at: http://gentleseas.blogspot.sg/2014/08/air-independent-propulsion-aip.html [Accessed Date: 17 October 2015].
39. Saab Kockums. 2010. "At the Leading Edge of Submarine Technology." Available at: http://saab.com/globalassets/commercial/naval/submarines-and-warships/submarines/a26/saab_kockums-a26_brochure_a4_final_aw_screen.pdf [Accessed Date: 17 October 2015].
40. Rex, Patrick. 2011. "Sea 1000: Conventional Air Independent Propulsion." *Asia-Pacific Defence Reporter* 37(10): 40–44.
41. Coates, Peter. 2014. "Air Independent Propulsion Technologies and Selection." *Submarine Matters*. Available at: http://gentleseas.blogspot.sg/2014/08/air-independent-propulsion-aip.html [Accessed Date: 17 October 2015].
42. Lakeman, JB and Browning, DJ. 2003. "The Role of Fuel Cells in the Supply of Silent Power for Operations in Littoral Waters." Paper presented at the RTO AVT Symposium on "Novel Vehicle Concepts and Emerging Vehicle Technologies," Brussels, Belgium, April 7.
43. Howaldtswerke-Deutsche Werft. 2011. "Fuel Cell AIP – Silent Power for Submarine Application." Available at: https://www.thyssenkrupp-marine systems.com/en/hdw-fuel-cell-aip-system.html [Accessed Date: 17 October 2015].
44. Hauschildt, Peter and Hammerschmidt, Albert. 2011. "PEM Fuel Cell Systems – An Attractive Energy Source for Submarines." Available at: http://info.industry.siemens.com/data/presse/docs/m1-isfb07033403e.pdf [Accessed Date: 14 June 2013].
45. Cavas, Christopher. 2014. "Israel's Deadliest Submarines are Nearly Ready." *Defense News Intercepts*. 14 August. Available at: http://intercepts.defense news.com/2014/08/israels-deadliest-submarines-are-nearly-ready/ [Accessed Date: 17 October 2015].
46. Coates, Peter. 2014. "Air Independent Propulsion Technologies and Selection." *Submarine Matters*. Available at: http://gentleseas.blogspot.sg/2014/08/air-independent-propulsion-aip.html [Accessed Date: 17 October 2015].
47. Mackenzie, Christina. 2012. "Spanish S-80 Subs Sailing Forward." *Aviation Week & Space Technology*. Available at: http://www.aviationweek.com/Article.aspx?id= /article-xml/AW_07_02_2012_p24-469557.xml [Accessed Date: 17 October 2015].
48. Gordon, Arthur. 2012. "Regional Surge in Submarines and Technologies." *Defence Review Asia*, October 22. Available at: http://www.defencereviewasia.com/articles/188/REGIONAL-SURGE-IN-SUBMARINES-TECHNOLOGIES [Accessed Date: 17 October 2015].
49. Kopp, Carlo. 2010. "Evolving ASW Sensor Technology." *Defence Today* 8(5): 26–29.
50. Kopp, Carlo. 2010. "Air Independent Propulsion – Now a Necessity." *Defence Today* 8(5): 10–12.

8

From Subsonic to Hypersonic Cruise Missiles: Revolution or Evolution in Land Attack Capabilities?

Kalyan M. Kemburi

Since the beginning of the 1990s, the advent and success of US "cruise-missile diplomacy"[1] has resulted in the emergence of cruise missiles as a coercive political tool and a versatile military weapon. During the initial phase of the last two decades, cruise missiles were predominantly deployed by a select group of advanced industrial countries, in particular the United States. However, the last decade had witnessed a new trend with emerging industrial countries showing increased propensity to develop cruise missiles. In addition to the effectiveness in carrying out all-weather precision strikes at standoff ranges, the rationale for acquiring cruise missiles is also driven due to the belief in the weapon systems' ability to penetrate air defenses and the affordability in development and deployment.[2]

The successful missions since the First Gulf War and the subsequent discourse has raised the profile of cruise missiles resulting in several military establishments initiating programs to develop land-attack cruise missiles (LACM). Within Asia, five countries—China, India, Pakistan, South Korea, and Taiwan—have active programs aimed at developing LACMs (most of them flying at subsonic speeds). In recent years, some of the Southeast Asian countries (Indonesia, Malaysia, and Vietnam) have expressed interest in procuring cruise missiles. Japan, which is constitutional constrained to develop offensive military capabilities, has also indicated interest in a weapon system that could endow the Japan Defense Agency with preemptive strike capabilities. For countries such as Japan, South Korea, and Taiwan, high-cost involved in deploying missile defenses and the normative and/or treaty restrictions of developing ballistic missiles, have made LACM an attractive system in achieving counter force options against the adversary's ballistic missiles and artillery systems.[3]

Most of the LACM deployed worldwide are characteristic of stealth, standoff range, and capability to carry all-weather precision strikes; this versatility of the weapon system is achieved by the air-breathing subsonic flight, INS/GPS guidance[4] with optical or multi-spectral seeker for terminal phase, medium to low radar cross-section (RCS), terrain-hugging ability, minimal infrared signature, and multi-platform launch.

As with any military technology, there is always a cyclic dynamic between defense and offense. Deployment of cruise missiles also have resulted in concomitant developments in defense: active-counter measures include advances in early warning systems and the deployment of AWACS (Airborne Warning and Control System) aircraft and radars based on aerostats as well as strengthening of passive defenses such as hardening of installations holding critical assets like aircraft or command and control equipment. Moreover, new operational requirements, especially the need to reduce sensor-to-shooter-to-target times have intensified efforts in the direction of high-speed air-breathing LACMs. Five countries in Asia—China, Japan, India, South Korea, and Taiwan—either have civilian or/and military programs aimed at developing supersonic and hypersonic air-breathing systems. Although a firm speed categorization is difficult, it is generally agreed that supersonic (powered by ramjet engine) operate in the range of Mach 2 to 4 and hypersonic (scramjet engine) over Mach 5.[5]

This chapter endeavors to evaluate the operational utility and technological feasibility of developing high-speed air-breathing propulsion systems for LACM. The first section brings forth the various operational opportunities and technological challenges associated in developing supersonic and hypersonic LACMs. The subsequent section includes an assessment of hypersonics and provides discussion on whether high-speed air-breathing missiles provide the military advantage commensurate with the resources invested to develop these systems. The chapter concludes that due to technological factors and operational opportunities offered by supersonics, over this decade LACM powered by supersonic engines would increasingly become an attractive option and feasible complement for the existing systems involved in generating firepower. The conclusion section also includes analysis on the impact of high-speed cruise missiles on military capabilities in the Asia-Pacific, and the concomitant implications for balance of power dynamics.

Hypersonic and supersonic cruise missiles

Rene Lorin of France first proposed the idea of a ramjet in 1913 and subsequently Albert Fonó, a Hungarian inventor, secured a German patent in 1934 for an earlier version of the ramjet engine. After the end of Second World War, research and conceptual thinking on high-speed air-breathing technology widened in scope and intensity; the initial developmental work was in advanced indusial countries of the era, the USSR, the United Kingdom, Germany, and the United States, who were later joined by Germany and France.

While ramjet engine was successfully developed and deployed in a wide range of missiles, scramjet development involved much more complexity and after many fits and starts, the first successful scramjet aircraft was successfully tested in 2004. The vehicle X-43A developed by NASA reached Mach 7 and completed the planned 10-second test.

This section includes a brief overview of the existing air-breathing propulsion systems (turbojet, turbofan, ramjet, and scramjet engines); comparative analysis of the operational utility of subsonic, supersonic, and hypersonic LACM; and finally discussion on the status of civilian and military programs in Asia aimed at developing supersonic and hypersonic technology.

Air-breathing propulsion systems

The propulsion system has a critical impact on determining the range, speed, and payload of a missile. The major propulsion systems currently used to propel air-breathing cruise missiles the world over are turbojet (French Storm Shadow LACM and Chinese C-802/ YJ-82 ASCM), turbofan (US Tomahawk LACM), and ramjet (Russian Sunburn ASCM and Russia-India BrahMos LACM).[6]

Turbojet engines have high thrust levels and can reach supersonic speed. However, the rate of fuel consumption limits turbojets to propel missiles to longer ranges, a limitation overcome by turbofan technology. Turbofan engines consume less fuel than turbojet engines of similar size, thereby increasing the payload and the range needed for deep strikes.[7]

Turbojet and turbofan propelled cruise missiles operate usually within the subsonic speeds. For higher speeds, missiles should be equipped with either ramjet or scramjet systems. Of the two systems, currently only ramjets are operational and scramjet technology is still under development and testing. Although heavier than their subsonic counterparts, these engines are simple in that they do not have major moving parts.

Both ramjet and scramjet engines require supersonic airflow to operate, therefore these engines need an external booster (either a rocket or aircraft) to take them to the "takeover velocity." The main difference between ramjet and scramjet is that in the former the combustion takes place at subsonic speeds and in the latter at supersonic speeds.[8]

In ramjet engines, the compression of air is achieved by decelerating the incoming supersonic air to subsonic at the inlet; this mechanism is effective and efficient up to Mach 4. For cruise speeds higher than Mach 4, combustion should take place at supersonic airflow. This is accomplished in scramjet engines. The main challenge in designing scramjets involves intake and combustor design, which have to withstand enormous heat.[9] The main reason for this enormous heat load is due to the high energy of the oncoming supersonic flow and high gas density from compression.[10] Therefore as noted by Richard Hallion (was a Senior Adviser for Air and Space Issues at the Pentagon), in designing hypersonic air-breathing engines, "aerodynamic heating considerations become at least as significant as concerns over aerodynamic design and structural design."[11]

Operational comparison between subsonic and super/hypersonic LACMs

Although the first reported operational deployment of cruise missiles was during the Second World War, it was only since the First Gulf War that cruise missiles started to assume a critical role in achieving tactical as well as strategic objectives on the battlefield. LACM missions could be broadly categorized into the following:

- Strategic-strike operations[12]
- Strikes against air bases as well as command and control (C^2) centers and support infrastructure
- Suppress and/or destroy air defenses and radar installations
- Counter-force operations against ballistic missile launchers and artillery units

In spite of the operational successes achieved by subsonic LACMs in last 20 years, the development of active-counter measures (including advances in early warning systems and the deployment of AWACS) as well as the emergence of new operational requirements, especially time critical targets, could potentially affect the operational effectiveness of cruise missiles flying at subsonic speeds. Supersonic and hypersonic missiles can overcome the constraints imposed by time, distance, and advanced early-warning and air-defense systems as well as shorten the shooter-to-target time, thereby holding multiple targets under threat. High speed LACMs have the potential to provide additional options for the following operations:

1. Time critical targets:

Most of the LACMs deployed worldwide fly at subsonic speeds of around 800km/h. Currently, with stealth and precision as priorities, these subsonic speeds are sufficient in achieving most of the current operational needs. However, time critical targets including missile launchers and mobile C^2 units require minimum sensor-to-shooter-to-target times.

The criticality of time and detection in countering ballistic missile operations is best illustrated by coalition efforts during the Operation Desert Storm in 1991. Despite devoting 20 percent of F-15E air sorties for 'Scud hunt', the coalition forces could not destroy even one Iraqi scud launcher.[13] Similarly, with actionable intelligence, high-speed missiles have a critical utility in targeting high-value targets; a lacuna highlighted by the 1998 failure of subsonic Tomahawks to arrive on time before Osama bin Laden could flee the targeted location in Afghanistan.

2. Modern air defenses:

Although low RCS and terrain-hugging flight path enables the subsonic cruise missile to evade air defenses, if detected the subsonic cruise missiles are highly susceptible to terminal defenses including anti-aircraft artillery and MANPADS. Moreover, with recent advances in radar technology subsonic cruise missiles, even if flying at low altitudes, could be detected by AWACS and aerostats and countered by aircraft equipped with look-down/shoot-down radar. Although LACM flying at supersonic or hypersonic speeds are relatively more easily detectable due to their high IR signature, their high speeds coupled with maneuverability make them a difficult target for air defenses in Asia.

3. Hardened targets and mountain warfare:

Because of their high-velocity impact, high-speed LACMs are also very useful as penetrators for targeting hardened buried targets. According to a report published for the US Air Force in 2000, a 250lb hypersonic penetrator can acquire the same penetration depth and impact as a 5,000lb gravity bomb.[14]

In mountain warfare, a manned aircraft has to operate within narrow maneuvering spaces with high ridgelines as well as being required to perform steep dives, which at times result in loss of altitude endangering the aircraft and crew. In this demanding environment, high-speed LACM with their ability to engage in powered climbs and dives offer a critical capability. As mentioned earlier, one of the most operationally significant attributes of cruise missiles is the flexible flight path that enables the missile to engage in a multi-directional attack on the target, which imposes additional geometric requirements on the defenses.[15] In addition to this multi-directional attack path, a ramjet-powered missile could also perform over a wide altitude bracket and can engage in powered climbs and dives, which would impose severe processing and cuing difficulties for the air defenses.

Supersonic and hypersonic technology in Asia

In Asia, China, India, Taiwan, Pakistan, and South Korea have active cruise missile programs; of these countries, presently only India and Taiwan have deployed supersonic LACM. On the civilian side, China, Japan, and India have programs aimed at developing hypersonic systems, especially for space access.

In spite of the research and technical capabilities to develop a ramjet engine, turbojet and turbofan engines currently propel LACMs in China. The only supersonic air-breathing missile in the Chinese inventory is the Russian derived Sunburn ramjet powered ASCM.[16] With the initiation of the indigenous production of this missile (designated Fu-Feng-1 (FF-1) or JL-9) by the China Aerospace Sciences and Industry Corporation, Beijing could potentially use a derivative of this ramjet engine to power a LACM. China Aerodynamics R&D Center and the National University of Defense Technology are currently working towards scramjet propulsion, pulse-detonation engines, and turbine-based combined cycle (TBCC) engines with an aim to eventually develop hypersonic missiles and aircraft. Further, the China Academy of Aerospace Aerodynamics has reportedly developed an experimental scramjet.[17] In January 2014, reports indicated that China had successfully tested a hypersonic system—a potential precursor for a high-speed cruise missile.[18]

The Japanese Aerospace Exploration Agency (JAXA) is involved in developing high-speed air-breathing propulsion for a hypersonic aircraft.[19] JAXA is also collaborating with institutions based in Australia, Germany, Italy, and the US in developing scramjet-based systems for space access. In 2012, Japan reportedly tested a rocket-based combined-cycle engine model under Mach 8.[20]

India currently deploys the ramjet powered supersonic LACM Brahmos flying at Mach 2.5–2.8.[21] In 2008, India and Russia agreed to fund jointly a kerosene-based hypersonic Brahmos 2 LACM with cruising speeds of Mach 5 to 8.[22] Concurrently, the Defense Research and Development Organization (DRDO) is working on a hypersonic system called HSTDV (hypersonic technology demonstrator vehicle), which could fly at Mach 6 to 7 propelled by scramjet. In 2012, *Aviation Week* cited an unnamed DRDO official who noted that the initial ground testing of the vehicle for nearly 20 seconds was successful and possibly a full flight test would be conducted by 2014.[23] For space access, India's civilian space agency—Indian Space Research Organization—has been working on a hydrogen-fueled scramjet engine. The aim is to develop a two-stage-to-orbit (TSTO) reusable launch vehicle propelled by a scramjet engine capable of placing a payload in low earth orbit; media reports indicate the first supersonic combustion ground tests were conducted in 2005.[24]

Taiwan's Hsiung Feng III (HF-3) LACM is propelled by a ramjet engine flying at a maximum speed of Mach 2 with an estimated range of 150 to 200 km. Initially developed as an ASCM by Chung Shan International Institute of Science and Technology, it was later reported that the missile also has land attack capabilities and entered into service in 2008.[25]

South Korea emerged as the most recent country in Asia to report a cruise missile program. In the last two years, media reports noted that Seoul has also been developing a supersonic Haeseong-2 LACM from the existing ASCM Haeseong-1 (Sea Star, or SSM-700K).[26] In September 2011, *Korea Times* reported that the missile is slated for deployment by the end of 2013 and has a range in excess of 500 km.[27] Additionally, the Korea Aerospace Research Institute (KARI) has on the drawing board a two-stage Mach 4 scramjet propelled surface to air interceptor. Reportedly KARI has ground tested various scramjet components required for this concept.[28]

Thus, as noted supersonic/hypersonic programs in Asia are gradually transitioning from the level of basic science and technology to testing phase for components and/or complete systems. Nevertheless, the programs require more substantial investments and at least another 10 to 15 years to move to the level of platforms and subsystems. As of now it seems the drive towards high-speed cruise missiles is relatively a technological pull rather than a push due to operational requirements. Compared to the military sector, supersonic/hypersonic programs in the civilian sector might receive more resources, especially with increased demand for affordable access to low earth orbit as well as for high-speed transportation. This progress in the civilian sector would provide further acceleration for supersonic/hypersonic cruise missiles—especially given it is less daunting to hit a target than to transport a payload.

Assessment

The quintessential question is whether high-speed air-breathing missiles provide the military advantage commensurate with the enormous technical and financial resources necessary to develop these systems. Before dwelling further in finding inputs for this question, it is pertinent to differentiate between the resources required and operational objectives fulfilled in developing and deploying ramjet (supersonic) and scramjet (hypersonic) propelled missiles respectively.

At the technological level, currently ramjet powered missiles are either deployed or the related propulsion and material technologies are in an advanced stage of development and testing, whereas technology required for scramjets has been in a state of "development" since the 1950s. The last two decades have witnessed the maturity of ramjet technology resulting in its wide-scale application in missiles of various configurations ranging from surface-to-surface, air-to-air, air-to-surface, anti-ship, and recently even for anti-tank systems. Over the next ten years, there is an immense scope for further refinement of this technology as well as adoption by more countries in Asia.

The section starts with a discussion of the technological challenges involved in developing scramjet engines. This discussion is followed by a comparative analysis of operational choices between supersonic and hypersonic missiles.

Technological challenges

Compared to turbojet and turbofan engines, ramjet and scramjet engines are much simpler in design as there are no rotating components (compressors or turbines); however, scramjet involves a more complex operating cycle. The point of ignition for scramjet engine is at an altitude of around 65,000 feet and requires air inflow at speeds of Mach 4.5, which necessitates a powerful booster. The complexity of this operating cycle does not end with reaching the takeover velocity and altitude, but only begins. In *Hypersonic Power Projection*, Richard Hallion provides a succinct description of the combustion processes involved:

> During the transition from rocket-lofting to scramjet ignition, "capturing" the airflow for proper internal flow, fuel-mixing, and ignition is complex, requiring efficient flight control and propulsion system.[29]

Two other factors also play a critical role in the performance of scramjet-powered vehicle: heat load and aerodynamic stability. The three tests of Boeing X-51 Wave Rider provided a preview to the scientific community the difficulty in overcoming these challenges. The first test in 2010 completed only 140 seconds of flight rather than the planned 300 seconds because hot gases burned through the seals between engine and nozzles. In the second test in 2011, after the separation from the booster, the scramjet did not transition from ethylene to the main fuel JP7 (hydrocarbon). During the third test in 2012, one of the four control fins required for stable aerodynamic flight malfunctioned.[30]

Additionally, the fuel and the material used for the missile fuselage influences the developmental, acquisition, and operating costs. The current high-speed missiles using ramjet engines are based on less complex, stable, and affordable fuels such as kerosene. Air-breathing missiles using hydrocarbon fuel with uncooled combustion chambers have a top speed of Mach 6, which can be increased to Mach 8 with endothermic cooling of the combustion chamber; for higher speeds, more exotic and expensive fuels are the order of the day.[31]

As noted in earlier, high-speed missiles could provide critical capability against hardened buried targets. However, an important point for consideration is that at supersonic speeds up to Mach 4 a steel penetrator retains its strength,[32] but at hypersonic speeds the penetrators should have a much harder casing such as tungsten (which might increase the cost of the weapon). Similarly, the current material used in missile fuselage loses its strength at hypersonic speeds necessitating use of tungsten-based nose caps, structures based on nickel alloys, and other special carbon-based material to withstand the enormous temperatures. The final point for consideration is that the impact of hypersonics on the current range of navigation and terminal guidance systems also requires further study.[33]

Operational opportunities

The scramjet-propelled missile provides enormous advantage in terms of greater range and reduced time to critical targets; however, limitation in C4ISR systems and rigid organizational structures of most Asian militaries limit the utility of hypersonics. This raises two interrelated questions: first, whether there is an immediate requirement for the militaries in Asia to consider hypersonic missiles. A negative response leads to the next question of whether supersonic cruise missiles are adequate for the current tactical and strategic objectives. The rest of the section aims to answer this issue through a comparative analysis of operational choices between supersonic and hypersonic missile:

First, critics highlight that the prevailing C4ISR systems in Asia might limit the operational advantages accrued from improvements in speed of the cruise missile. As noted earlier, one of the critical missions for high-speed LACMs is to engage in counter-force operations targeting adversary's missile and artillery units. A study conducted for the US Air Force in 2000 provides a hypothetical timeline of 8 minutes for a theatre ballistic missile launcher to fire its missile. This provides approximately 4 minutes for target detection, recognition, and identification as well as for the decision-making processes, which ranges from assignment of weapon, mission planning, to actual strike; and the remaining 4 minutes for the flight of the counter strike missile.[34] For most militaries in Asia, 4 minutes for search and identification of the missile launcher itself is a challenging task, let alone including the decision making process in this 4-minute loop.

Second, intermediate ranges in the international arena are strategic in nature for Asia. Unlike the United States, which requires global strike capabilities with reduced flight time, for major militaries in Asia most of their targets are within 1,500km range. As mentioned in Table 8.1, if a supersonic LACM requires around 17 minutes to reach a target at 1,000kms, a hypersonic missile reaches the target in less than 10 minutes—a difference of approximately 8 minutes. Is 8 minutes critical for military outcome? Even if this timeline is critical, as mentioned earlier most Asian militaries are neither equipped with necessary C4ISR systems nor are the civilian and the military organizational structures geared to respond in a time critical manner.

Third, progress in installing advanced early warning and air defense systems is either slow or limited in Asia, with the exception being Japan. Moreover, even if the present early warning systems aided by aerostats and AWACS are able to detect the incoming subsonic cruise missiles, a concerted attack by subsonic and supersonic LACM together with theatre ballistic missiles would create processing difficulties for any advanced early warning system, especially because of different flight trajectories and speeds of these three missiles. These processing difficulties could range from failure to detect all the incoming missiles to friendly casualties.

Conclusion

As noted by Andrew Krepinevich, "advances in technology typically underwrite a military revolution, they alone do not constitute the revolution."[35] Accordingly, this chapter concludes that given the limitations in C4ISR capabilities, rigid organizational and decision-making

Table 8.1 Time-to-target estimates for various cruise missile types

Speed of missile / Distance to the target	Subsonic (800 km/h.)	Supersonic (Mach 2.8)	Hypersonic (Mach 6)	Time difference between Mach 2.8 and 6
300 km	22 min 30 sec	5 min 17 sec	2 min 57 sec	2 min 20 sec
500 km	37 min 30 sec	8 min 49 sec	4 min 55 sec	3 min 54 sec
1,000 km	75 min	17 min 38 sec	9 min 30 sec	8 min 8 sec

Note: Mach 1 = 1,225 km/h.; Mach 2.8 = 3,400 km/h.; Mach 6 = 7,300 km/h. As the only operational supersonic LACM is Brahmos 1, for the purpose of this chapter the author has based the calculations on Brahmos flight speeds. Brahmos 1 has a speed of Mach 2.8 and Brahmos 2 (under development), the hypersonic version, aims to fly at Mach 6.

processes, and enormous resources necessary to deploy a hypersonic missile, over the next 10 to 15 years supersonic LACM offer a more viable complement to the existing cruise and ballistic missiles. The hypersonic air-breathing missile is a key emerging technology; however, for an effective and efficient use of this technology concomitant changes are necessary in organizational structures, decision-making processes, operational concepts, and C4ISR systems. The following section further elaborates the implications of high-speed cruise missiles on regional security dynamics, particularly issues such as anti-access/ area denial, counter-force options, and asymmetrical means against a more advanced adversary and briefly discusses the emerging operational and organizational issues in this regard.[36]

First, regional security implications:

- Anti-access/area-denial (A2/AD): With an ability to penetrate defenses and strike with precision, perform under all-weather conditions along with a relative affordability in developing and deploying, a subsonic cruise missile is one of most potent A2/AD weapon. A high-speed cruise missile takes the A2/AD warfare to the next level. A high-speed cruise missile has the potential to change the contours of the A2/AD warfare by severely restricting area access and maneuverability of the intervening forces. To illustrate, a launch platform carrying a standard subsonic cruise missile flying for 10 minutes at Mach 0.7 could strike targets at a distance of 90 miles, and hold an area of around 20,000 mi^2 at risk. Whereas at Mach 5, the missile can strike targets at 575 miles and holds an area of approximately 1 million mi^2 at risk. For comparison, one of the crisis hot spots in Asia, the South China Sea is 1.1 million mi^2.

- Offense is cheaper than defense: For most countries—especially Japan, South Korea, and Taiwan in the Asian context—the high cost involved in deploying missile defenses and the normative and/or treaty restrictions of developing ballistic missiles have made cruise missiles an attractive system in achieving counter-force options against the adversary's ballistic missiles and artillery systems.[37] For example, in a hypothetical conflict across the Taiwan Strait, a high-speed LACM on the recently inducted Tuo Chiang corvette provides near real-time quick reaction strike options for Taipei against China's mobile ballistic missile and ASCM launchers, albeit having escalatory consequences.
- Over the next 15 years, only four countries—China, Japan, India, and South Korea—in Asia have the potential to develop and deploy supersonic LACMs. However, with dissemination of ideas, diffusion of technologies, and a "buyer's" weapons market, relatively more countries could have access to advanced LACMs. With induction of limited number of supersonic LACMs—even in the absence of a complete C4ISR package—these countries acquire asymmetrical capabilities to tie down larger and more sophisticated military forces. Therefore, high-speed cruise missiles could potentially provide "offsetting" alternatives to compete asymmetrically with a more technologically advanced adversary.

Second, in Asia, a supersonic LACM would become an attractive option due to the following factors:

- Reduces sensor-to-shooter-to-target times: a supersonic LACM flying towards a target at 1,000 km has clear time advantage of almost 58 minutes over subsonic LACM.
- The kinetic energy of a supersonic missile not only increases the explosive power of a warhead but also facilitates reduction of the warhead payload, which helps in expanding the range of the missile.
- Supersonic LACMs used in conjunction with subsonic and theatre ballistic missiles create processing difficulties for any advanced early warning system.

Third, as much as new technology creates organizational efficiency and effectiveness as well as new means to fight, it also possibly creates concomitant unintended consequences.

- One such unintended effect could be a new culture of micromanagement by the senior leadership. With C4ISR systems providing a near real-time picture of the battlefield along with the ability to pick and choose the targets, there is a danger of generals becoming tacticians. For example, during the Vietnam War, the induction of relatively new technology of helicopters created an unintended effect of senior commanders hovering over the battlefield to manage the tactics, transforming into "squad leaders in the sky."[38]

- The day after the attack: with high-speed precision strikes, the time required to realize the target list would be substantial reduced. Therefore, it becomes imperative for the political and military leadership not just to know how to conduct a blitzkrieg but what to do after the blitzkrieg.

Finally, would the induction of high-speed cruise missiles be an evolutionary or a revolutionary phenomenon? In an ever-evolving security environment with diffused military capabilities, reliance on a single weapon system or a unidirectional policy initiative creates an illusion of success and fails to produce the required political objectives. A new weapon system requires associated operational innovations and at times even organizational changes as well as upgrades in support infrastructure and systems. Asian militaries are still in the process of inducting significant number of subsonic LACMs—and supersonic LACMs in some cases—as well as currently working on innovative concepts and organizational changes that aim to take advantage of these systems in affecting the outcomes on the battlefield; therefore, induction of high-speed missiles is evolutionary.

Notes

Kalyan M. Kemburi is an Associate Research Fellow with the S. Rajaratnam School of International Studies, Nanyang Technological University, Singapore. He was previously with the James Martin Center for Nonproliferation Studies, California, and the UN Office for Disarmament Affairs, New York.

1. David Tanks, *Assessing the Cruise Missile Puzzle: How Great a Defense Challenge*, Boston, MA: The Institute for Foreign Policy Analysis, October 2000, p. 7.
2. In this regard, two recent incidents have attested to the effectiveness and survivability of cruise missiles: First, the failure of coalition missile defenses against the Iraqi cruise missile attacks during the 2003 Operation Iraqi Freedom (OIF). Second, Hezbollah's successful attack against an Israeli naval vessel in 2006 with an anti-ship cruise missile (ASCM).
3. Dennis M. Gormley, *Missile Contagion: Cruise Missile Proliferation and the Threat to International Security*, Annapolis, MD: Naval Institute Press, 15 September 2010.

4. To increase precision and safeguard against counter measures, in addition to Inertial Navigation System (INS) and GPS, some LACM also use either Digital Scene-Mapping Area Correlator (DSMAC) or Terrain Contour Matching (TERCOM).

5. SCRAMJET: Supersonic Combustion Ramjet.

6. Tanks, *Assessing the Cruise Missile Puzzle*, pp. A 2–3.

7. Tanks, *Assessing the Cruise Missile Puzzle*, pp. A 2–3.

8. William H. Heiser, David T. Pratt, Daniel H. Daley, and Unmeel B. Mehta, *Hypersonic Airbreathing Propulsion*, AIAA Education Series, Washington DC.: American Institute of Aeronautics and Astronautics, 1994, pp. 22–24; and V. Babu, *Aircraft Propulsion*, Boca Raton, FL: CRC Press (Taylor and Francis), 2009, pp. 191–192.

9. Heiser, Pratt et al., *Hypersonic Airbreathing Propulsion*, pp. 22–24; and Babu, *Aircraft Propulsion*, pp. 191–192.

10. Heiser, Pratt et al., *Hypersonic Airbreathing Propulsion*, p. 24.

11. Richard P Hallion, *Hypersonic Power Projection*, Mitchell Paper 6, Virginia: Mitchell Institute for Airpower Studies, June 2010, p. 9.

12. These operations include "disabling the enemy's center of gravity [that] will result in the loss of his ability or will to offer further resistance to friendly forces in achieving their strategic objectives" [center of gravity is defined as key military, economic, or political assets]. For the purposes of this chapter, these operations are discussed in the context of strikes carried by conventional weapons against non-nuclear targets/key nodes in the local theater of conflict that provide strategic gains. For more information, refer Andrew F. Krepinevich and Robert C. Martinage, *The Transformation of Strategic-Strike Operations*, Washington DC: Center for Strategic and Budgetary Assessments, March 2001, p. 1.

13. Dennis M. Gormley, *Dealing with the Threat of Cruise Missiles*, Adelphi series 339, Abingdon, Oxon: Routledge, p. 64.

14. Ronald P. Fuchs, Armand J. Chaput, David E. Frost, Tom McMahan, David L. Vesely, David A. Deptula, Douglas L. Amon, Alan D. Bernard, Frederick S. Billig, Leonard F. Buchanan, Ramon L. Chase, Natalie W. Crawford, Thomas A. Cruse, Darryl P. Greenwood, Richard Hallion, Susan E. Hastings, Daniel T. Heale, David Jablonski, John E. Jaquish, Ray O. Johnson, O'Dean P. Judd, Ann R. Karagozian, Sherman N. Mullin, Matthew P. Murdough, George F. Orton, Vincent L. Rausch, Howard K. Schue, Jason L. Speyer, David M. Van Wie, and Michael I. Yarymovych, "Report on Why and Whither Hypersonics research in the US Air Force," p. 55.

15. Richard K Betts, ed., *Cruise Missiles: Technology, Strategy, Politics*, Washington DC: The Brookings Institution, 1981, p. 81.

16. "Fu-Feng-1/JL-9 (SS-N-22 'Sunburn')," *Jane's Strategic Weapon Systems*, December 5, 2012.

17. Unmeel Mehta, "Hypersonic Technologies and Aerospace Plane," *Aerospace America*, December 2008; in Lexis-Nexis Academic, "China's Scramjet Ambitions," *Aviation Week & Space Technology*, 3 September 2007; in Lexis-Nexis Academic and Craig Covault, "China Accelerating Scramjet Development," *Aerospace Daily & Defense Report*, 4 September 2007; in Lexis-Nexis Academic.

18. Ankit Panda, "China Tests Hypersonic Missile Vehicle," *The Diplomat*, 14 January 2014, http://thediplomat.com/2014/01/china-tests-hypersonic-missile-vehicle/, accessed on 20 May 2013.

19. Mehta, "Hypersonic Technologies and Aerospace Plane."

20. Foluso Ladeinde and Jeff Dalton, "High-speed Air-breathing Propulsion," *Aerospace America*, December 2012, in Lexis-Nexis Academic.

21. "BrahMos (PJ-10), 3M55 Yakhont," *Jane's Air-Launched Weapons*, 28 March 2011, Martin Sieff, "BrahMos-2 Tests Mark Major Progress On Indian Cruise Missile," *Space Daily*, 6 March 2008, http://www.spacedaily.com/reports/ BrahMos-2_Tests_Mark_Major_Progress_On_Indian_Cruise_Missile_999.html, accessed on 20 May 2013; Sayan Majumdar, "The Brahmos Punch," India Defence, 10 May 2005, http://indiadefence.com/brahmospunch.htm, accessed on 20 May 2013, and T.S. Subramanian, "BrahMos Success," *Frontline*, Vol. 20, No. 05, 01–14 March 2003, http://www.flonnet.com/fl 2005/stories/20030314002509400.htm, accessed on 15 May 2013.

22. "Cruise Control," *Aviation Week & Space Technology*, 29 August 2011; in Lexis-Nexis Academic and "Cruise Control," *Aviation Week & Space Technology*, 13 September 2010; in Lexis-Nexis Academic. Additionally, reports also indicate India is developing a second ramjet-powered LACM with range up to 1,000km and potentially flying at Mach 3.2.

23. According to Dr. V.K. Saraswat, Director General of DRDO, through this project the organization aims to "demonstrate the performance of a scramjet engine at an altitude of 15 km to 20 km. Under this project, we (DRDO) are developing a hypersonic vehicle that will be powered by a scramjet engine. This is dual-use technology, which when developed, will have multiple civilian applications. It can be used for launching satellites at low cost. It will also be available for long-range cruise missiles of the future." For more information in this regard, refer T.S. Subramanian, "DRDO developing hypersonic missile," *The Hindu*, 09 May 2008, http://www.thehindu. com/todays-paper/tp-national/article1254728.ece, accessed on 17 October 2015 and Jay Menon, "Homegrown Hypersonics," *Aviation Week & Space Technology*, 26 November 2012; in Lexis-Nexis Academic.

24. "ISRO Achieves Breakthrough in Supersonic Combustion," *ISRO Newsletter*, October 2005–March 2006, http://www.isro.org/newsletters/scripts/ newslettersin.aspx?ISROachievesOM56, accessed on 20 May 2013 and "Successful flight-testing of advanced sounding rocket," *ISRO Newsletter*, 3 March 2010, http://www.isro.org/pressrelease/scripts/pressreleasein.aspx? Mar03_2010, accessed on 10 March 2013.

25. Francis Leithen, "Taiwan Touts Its Aircraft 'Carrier Killer'," *Aviation Week & Space Technology*, Vol. 173, No. 29, 15 August 2011 and Michael Cole, "Update: Taiwan depicts HF-3 as a 'carrier killer'," *Jane's Defence Weekly*, 11 August 2011.

26. Bradley Perrett, "South Korea Works On New Missile Technology," *Aviation Week*, 01 June 2012, http://www.aviationweek.com/Article.aspx?id= / article-xml/DT_06_01_2012_p18-458092.xml, accessed on 20 May 2013 and Sebastien Falletti, "South Korea 'developing supersonic cruise missile'," *Jane's Defence Weekly*, 28 September 2011.

27. Further, the newspaper cited an unsubstantiated report from WikiLeaks, noting that South Korea has conducted around 10 tests of the Haeseong II between September 2007 and November 2009. For more information, refer Lee Tae-hoon, "Seoul Develops Supersonic Cruise Missile," *Korea Times*, 26 September 2011.

28. Guy Norris, "Scramjet Scramble," Aviation *Week & Space Technology*, 17 January 2011, Lexis Nexis.
29. Hallion, *Hypersonic Power Projection*, p. 9.
30. Graham Warwick, "Learning at Hyperspeed," *Aviation Week & Space Technology*, 26 November 2012; in Lexis-Nexis Academic.
31. Hallion, *Hypersonic Power Projection*, p. 27 and Fuchs et al., "Report on Why and Whither Hypersonics Research in the US Air Force," pp. 44–50.
32. Hallion, *Hypersonic Power Projection*, p. 26.
33. Currently LACM use INS/GPS for navigation, aided in some cases by TERCOM, and for terminal guidance multi-spectral seekers and DSMAC are some of the options; it is relevant to understand the ability of these systems to withstand the high stresses involved with hypersonic flight.
34. Fuchs et al., "Report on Why and Whither Hypersonics Research in the US Air Force," pp. 44–50.
35. Andrew F. Krepinevich, "Cavalry to Computer: The Pattern of Military revolution," *The National Interest*, Fall 1993/1994.
36. Unless noted otherwise, information for this section is sourced from Kalyan M. Kemburi, *Diffusion of High-Speed Cruise Missiles in Asia: Strategic and Operational Implications*, RSIS Policy Brief, 9 December 2014.
37. Gormley, *Missile Contagion: Cruise Missile Proliferation and the Threat to International Security*.
38. Thomas E. Ricks, *The Generals: American Military Command from World War II to Today*, New York: The Penguin Press, 30 October 2012, pp. 282–284.

9
The Potential Military Impact of Emerging Technologies in the Asia-Pacific Region: A Focus on Cyber Capabilities

Caitríona H. Heinl

This chapter explores why cyber capabilities are critical technological innovations that affect military capabilities and how they impact balances of power in the Asia-Pacific region. In this context, emerging technologies are defined as new types of research and development breakthroughs and know-how that could have a significant, altering effect on how militaries fight and gain advantages over rivals.[1]

The Asia-Pacific is a diverse region comprising countries that are at very different stages in terms of cyber technologies as well as strategy development and implementation. Capabilities that are considered new and emerging technologies in one state or a number of states are not always new throughout the region. However, given the current speed of technological change, most countries are still challenged by rapid developments, evolving cyber technologies, and increasingly complex cyber incidents. Moreover, this often means that timely and effective implementation of policy and legislation is made more difficult, especially since policy makers may not always fully understand the implications of new technologies quickly enough. Consequently, given several factors unique to this region such as recent tensions in the South China and East China Seas, existing territorial disputes, the uncertainty surrounding China as a regional military power and the United States' "pivot" towards Asia, as well as heightened concerns over North Korea, careful analysis of cyber capabilities and their possible impact is valuable.

Cyber attack is considered as a tier one threat to national security under the United Kingdom's 2010 national security strategy. The United

States also considers cyber threats as one of the most serious national security challenges under its 2010 national security strategy and in 2011 the US Department of Defense Strategy for Operating in Cyberspace designated cyberspace as an operational domain, in addition to the domains of land, air, maritime, and space.[2] However, prior to the discovery of Stuxnet in 2010, there was significant uncertainty in the public domain about the real nature of a "cyber weapon," in other words a cyber capability that could destroy hardware and cause kinetic damage. Stuxnet is argued to be the first known advanced cyber capability.[3]

Such advanced cyber capabilities can cause physical damage as opposed to virtual damage from incidents like distributed denial of service attacks and intellectual property or government data theft. These capabilities can also operate autonomously which means that special capabilities are not necessarily needed to use such a weapon, and although they are often identified with the Internet, an Internet connection is not essential.[4] In the case of Stuxnet, for instance, there was apparently no online dialogue once it was deployed.[5] In spite of this, once released, cyber capabilities can be reverse-engineered, tampered with, and used against new targets including the original creators.[6] This is particularly significant given that states can then fill gaps in their knowledge and technological capabilities.

Military structures are currently adapting to these emerging technologies and new realities. By mid-2012, more than 30 countries had or were in the process of creating military cyber units.[7] And official reports, such as the 2012 European Parliament Committee report, now recommend that designated cybersecurity and cyber defense units should be created within military structures.[8] Many therefore envisage that all future warfare will comprise a cyber component, in other words what has come to be known as "cybered conflict."[9] Cyber technologies may be used to either support or perhaps even substitute conventional war and it is possible that they could be used to destroy another nation's ability to wage war. Nevertheless, at this juncture, official reports so far suggest that most state and non-state actors do not yet have the technical expertise and operational sophistication for devastating attacks, including the ability to cause physical damage or overcome manual overrides.[10] Nonetheless, there is still a chance that unexpected significant outcomes could be caused by mistake by isolated state or non-state actors when using less sophisticated cyber attacks against poorly protected networks that control core functions.[11]

However, there is still significant debate on what in fact constitutes a cyber attack. Under the Tallinn Manual, it is defined by the manual's international group of experts as a cyber operation, whether offensive or defensive, that is reasonably expected to cause injury or death to persons or damage or destruction to objects (the manual which was published in 2013 is a non-binding, unofficial document that applies existing law to cyber warfare).[12] Examples of attacks under this definition include launching a cyber operation against a state's critical infrastructure or targeting command and control systems as opposed to traditional electronic warfare attacks like jamming.[13] The manual affirms that, generally speaking, the law of armed conflict applies to the targeting of any person or object during armed conflict irrespective of the means or methods of warfare employed. While cyber operations are not explicitly referred to in the existing law of armed conflict treaties, the ICJ previously affirmed that the principles of humanitarian law apply to all forms of warfare and all kinds of weapons, those of the past, present, and future.[14] The manual further states that principles regarding cyber attacks apply equally to situations in which cyber means are used to take control of enemy weapons and weapon systems, as in the case of taking control of an unmanned combat aerial system and using it to conduct attacks. Moreover, cyber operations may be an integral part of a wider operation that constitutes an attack and the law of armed conflict applies fully to such cyber operations.

Extensive discussion was also held as to whether interference by cyber means with the functionality of an object constitutes damage or destruction, including where it necessitates either data restoration or physical restoration of components, but no consensus was reached on this point. However, a cyber operation that results in large-scale adverse consequences such as blocking email communications throughout a country rather than damaging the system is not characterized as an attack by the international group of experts.

The means of cyber warfare are defined as cyber weapons and their associated cyber systems that are by design, use, or intended use capable of causing injury, death, or damage to or destruction of objects, in other words causing the consequences required for a cyber operation to qualify as an attack.[15] This includes any cyber device, material, instrument, mechanism, equipment, or software. The methods of cyber warfare are the cyber tactics, techniques, and procedures by which hostilities are conducted, in other words how cyber operations are mounted, as opposed to the instruments used to conduct them. For example, in an operation using a botnet to conduct a DDoS attack, the botnet is the means of cyber warfare while the DDoS attack is the method of cyber warfare. Active cyber defenses are included under the notion of methods of cyber warfare whereas passive cyber defenses are not, and cyber operations that qualify as ruses of war, in other words acts intended to mislead the enemy, are permitted. This includes dummy computer systems, transmission of false information, use of false computer networks, and psychological warfare activities.

Some estimate that the costs of acquiring offensive cyber capabilities for military targets are close to one billion US$ while critical infrastructure and civilian targets such as power plants and water plants are approximately 100 million US$.[16] This latter figure is affordable for most states and by these estimates it could take two to three years to develop an offensive cyber program. Furthermore, while many argue that advanced cyber capabilities are extremely difficult to create and that they require the resources of a nation state (or nation states), this it seems may not always be the case.[17] In the case of Stuxnet, for example, several security analysts argue that vast resources were in fact needed because its creators did not want the weapon or their identity to be discovered and high levels of intelligence were required whereas in future, attackers may instead prefer to be identifiable.[18] By contrast, bolstering defenses could be more difficult, more expensive and take longer over a period of approximately five to ten years.

Consequently, acquiring offensive or advanced cyber capabilities could seem financially attractive, in particular for less wealthier states in the region, relative to the higher costs of other weapons. Nevertheless, strong defense infrastructures may not be in place within these countries. This is particularly significant in this region where states are spending increasing amounts on arms capabilities and the military as compared to the US and EU, for example, where military budgets are being significantly reduced. Although, in relative terms, the overall US budget still far outweighs that of other countries, and it has not decreased its cyber defense budget to the same extent as other fields since it considers cyber as a national security priority. However, over the medium term, this balance could possibly shift, especially if defense reports are correct in their analysis that if China can maintain defense spending at its current levels as its economy continues to grow, the level of its defense spending may be three times as much as the US within the next 20 years.[19]

Non-state actors also further complicate this space since it does not solely concern state actors. Other actors such as cyber criminals, terrorists, hackers, hacktivists, and proxy actors engaged or supported by government, must be considered too. Moreover, some argue that growing cybercrime in this region could cause further instability because of its connections to espionage and military activities.[20] These points align with projections that the character of war is likely to continue to be shaped not only by a system of rival states but by forces outside the state-centric systems.[21] However, activities below the level of a "use of force" as understood in *jus ad bellum*, like cyber criminality, are not addressed here. This is because although cyber espionage, IP theft, and many criminal activities in cyberspace pose serious threats to states, corporations, and individuals, the application of international law on uses of force and armed conflict is not applicable to these threats in the physical world and therefore not applicable in the cyber domain according to the Tallinn Manual.[22] Furthermore, the Tallinn Manual considers that non-state actors' conduct may only be attributable to a state and give rise to its international legal responsibility if these actors are acting on the state's instructions, or under the state's direction or control in carrying out the conduct.

This is significant since states might contract with a private company to conduct cyber operations. Likewise, it is relevant when states have allegedly called upon private citizens to conduct cyber operations, including "cyber volunteers." However, this is not applicable if private citizens like hacktivists or patriotic hackers conduct cyber operations on their own initiative. Merely encouraging or otherwise expressing support for the independent acts of non-state actors does not seem to make a state responsible.[23] This is noteworthy given that in this region, there is a large and active "Internet militia" comprising hacker communities and information security experts, the majority of which, according to defense reports, are likely to be part of government structures or programs.[24] While others, although operating independently, are under the influence of or tolerated by national authorities.[25] These reports further assert that countries are likely to mobilize these groups as part of coordinated national efforts during periods of conflict.[26]

Training capable personnel is also a near-term challenge for most countries in the region since they are suffering from a skills shortage in this domain. The US Department of Defense's (DoD) is focusing on creating expertise by building a pool of talented civilian and military personnel, and developing Reserve and National Guard cyber capabilities for greater capacity. It also intends to foster rapid innovation as well as enhance its acquisition processes, in other words investing in people, technology, and R&D to build and sustain cyber capabilities. China and India on the other hand also have excellent information and communications technology (ICT) sectors but in addition to financial resources, they have large and growing pools of highly-skilled ICT professionals.

They will continue to invest in this domain as well as cultivate indigenous ICT to reduce reliance on imports and alleviate concerns over supply chain security.

In addition, cyber technologies are a good example of technologies with defense implications that arise from both the military and outside the military, therefore bringing its own set of unique challenges in that militaries must now work closely with the private sector. Lastly, effective acquisitions processes can be challenging given time constraints and the importance of ensuring a secure supply chain.

Regional balances of power and status of cyber capabilities and strategies

Accurate attribution of responsibility for a cyber incident can be difficult, which means that misunderstandings could possibly arise between states or tensions could escalate. Moreover, this often means it could be harder to ensure effective deterrence. The advantages of once-off and lower costs plus difficulties in appropriating blame can seem attractive, in particular for states with limited financial resources, capabilities, and expertise. And since non-state actors must be considered, as Keith Alexander, the former head of the National Security Agency in the US, previously claimed, cyber deterrence could be more difficult than nuclear deterrence.[27] Although if found responsible, states could risk reputation damage and physical retaliation. The US, for example, reserves the right to respond to cyber attack through kinetic means.[28] Furthermore, the US has recently issued statements of deterrence claiming that anonymity is not necessarily guaranteed and attribution can be possible. This is particularly relevant since strategists argue that asymmetry is assured when a conventional superpower like the United States goes to war because it is the world's strongest conventional power that does not have a peer competitor yet.[29] For example, in asymmetric warfare, the weaker side recognizes its enemy's military superiority and will avoid open confrontation while seeking instead to attain victory by adopting novel compensatory methods.[30] Therefore, in cyber, weaker entities can avoid this open confrontation and instead exploit vulnerabilities.

Recently, state actors' role in cyber matters and the use of cyber capabilities has increased significantly. Future projections expect that states and militaries will become more frequent sources and victims of cyber threats.[31] However, given the current sensitivities surrounding cybersecurity, in particular cyber capabilities, it can often be difficult to precisely ascertain the extent to which state actors in the Asia-Pacific region have developed or acquired capabilities, especially advanced cyber capabilities. In spite of this, increased military developments are expected within the region in cyber.[32] An April 2014 report on the Asia-Pacific region finds that the US, China, Australia, Singapore, and South Korea are leading the way in military aspects of cyber capabilities.[33] The report specifically highlights North Korea's increased use of cyber capabilities over the period 2013–2014 as especially concerning since it places the South Korean government under further pressure to ensure incidents do not escalate. South Korea is mainly concerned over North Korea's cyber capabilities, including cyber espionage and hacking, and its recent defense reform plan directs the defense ministry to bolster the country's cyber defensive capabilities against possible cyber attacks from North Korea.[34] Appendix 1 provides a summary of the status of military cyber issues across the Asia-Pacific region.

In particular, two major Asia-Pacific powers, the United States and China, are establishing themselves as the leading advanced cyber actors but mistrust has been high between counterparts.[35] These states are disputing the nature of international governance structures and the understanding of key terminology such as cyber/information warfare, cybersecurity/network, and information security. This relationship between these two states is important to regional security and it is watched closely by other Asia-Pacific countries.[36] And according to defense officials, the cyberspace domain is still being molded and contested, and it will continue to be challenged.[37] International negotiators additionally argue that diplomatic dialogue still needs to catch up with military developments in cyberspace.[38] For instance, the scope and manner of international law's applicability to cyber operations, whether in offence or defense, has been hitherto unsettled, there are many competing views, and terminology has been a particular challenge since many words have common usage but they also have specific military or legal meanings.[39] Furthermore, defense reports suggest that the scale and pace of the innovative and industrial capacity of countries like India and China will outpace many Western nations in a matter of years with China likely to attain and sustain global leadership in a number of technical areas including computer science.[40]

The US Office of the National Counterintelligence Executive describes Chinese actors as the world's most prolific perpetrators of economic espionage. And the US asserts that political and military spying can continue but China should stop economic espionage.[41] China's Defense Ministry, however, more recently claims that the government and military has never been involved in the theft of commercial secrets.[42]

Further, while Chinese cyber espionage is considered a real threat, cyber espionage capabilities are arguably different to those for strategic military operations.[43] In China, "the intelligence services also collect science and technology information to support government goals and economic advantage, while Chinese industry prioritizes domestically manufactured products for its technology needs."[44]

The United States is particularly anxious that foreign nations are exploiting unclassified and classified networks, that foreign intelligence organizations have the capacity to disrupt parts of its information infrastructure, and that there may be undetected malicious activities on DoD networks and systems. Foreign operations against public and private sector systems are growing in number and complexity, and defense networks are probed millions of times daily. Successful infiltrations have apparently led to the loss of thousands of files. Major concerns include the theft or exploitation of data, disruption or denial of access or service, destructive action including corruption, manipulation, or direct activity that threatens to destroy or degrade networks or connected systems. Moreover, vulnerabilities are perhaps higher in the US than some other countries in the region where systems are often less connected. Further, most ICT products are manufactured outside the country and this can mean there are higher risks in the supply chain at stages of design, manufacturing, service, distribution and disposal. By contrast, although China worries about US control in cyberspace, some argue that it is also particularly concerned over a local war, especially in the Taiwan region, as opposed to destroying critical information infrastructure in the US.[45] It therefore identifies US logistics, command and control, and C4ISR systems,[46] as well as systems such as US Pacific and Transportation Command as possible targets in a conflict over Taiwan, for example.[47] Likewise, China is also anxious about the internal control of potential subversives. It therefore co-opts spies because it considers that they too are targets that could turn against the regime.[48] However, while the Chinese military seems to have a fully realized doctrinal understanding of cyber warfare, some argue that it does not yet have the knowledge and technology to implement it. Furthermore, there could be a risk of exaggerating the PLA's capabilities, technology developments, and its infamous government-sponsored "patriot hackers" (IT/technical students are often automatically considered part of China's defense organs under the National Defense Reserve Forces program and drafted for training to reinforce this status).[49] Nonetheless, it is often difficult to fully know China's strategic intentions and it is habitually more ambiguous than other states.[50]

It could be destabilizing where one country considers an action permissible but another considers the same action as an act of war.[51] This is particularly concerning given that it is extremely difficult to control the production of cyber technologies, and agreement on key cyber terminology as well as the applicability of international law varies significantly between jurisdictions.[52] China for instance previously argued that a new treaty is required for use of these arms and it formerly submitted an "International code of conduct for information security" to the UN with Russia, Tajikistan, and Uzbekistan as a basis for a future resolution.[53] It also previously asserted that cyberspace should not become a new battlefield and states should prevent an arms race in cyberspace.[54] However, defense analyses predict that many South Asian states will undertake state-sponsored cyber programs facilitated by low barriers of entry, the availability of large pools of skilled manpower and extensive IT infrastructures.[55] While countries like China will be the most sophisticated players in the region out to 2040, other countries will still develop cyberspace activities in order to attempt to project influence which could be otherwise limited using only conventional instruments.[56]

And while US Intelligence Community reports calculate that advanced cyber state actors are unlikely to launch a devastating attack, this is based on the premise that there is no military conflict or crisis threatening vital interests.[57] Activities will therefore include espionage, interfering with digital information and networks, disrupting civilian and military infrastructure, manipulating financial systems and social networks, and constraining situational awareness.[58] These are especially important points considering the high national security sensitivities within this region and more recent military buildup, in addition to the friction between China and the United States over dominance in the region, conflicting ideologies, and current tensions over territorial disputes such as in the South China and East China Seas.

To conclude, maintaining stability in this region is of primary importance. To date and for the near term, it is unlikely that cyber capabilities will significantly change the power balance. Moreover, although these capabilities might be attractive to certain states, many militaries are still working on fully understanding the implications of these capabilities. Nevertheless, there is without doubt, a great need in the Asia-Pacific region for enhanced confidence-building and transparency measures in this space such as military-to-military engagements, dialogue, information sharing, joint exercises, official points of contact, and crisis communication procedures to prevent miscalculations, misunderstandings, false attribution, or escalation in tensions.

Appendix 1: Overview—Military Role in Cyber-Related Matters in the Asia Pacific.

Australia	The Department of Defense maintains sophisticated cybersecurity capabilities. The Australian Signals Directorate (ASD) is responsible for the development of signals intelligence capability. ASD is the Commonwealth Information Security Authority and maintains the Information security manual for Australian Government agencies. It runs the Cyber Security Operations Centre, which is responsible for defending against threats to Australian interests in cyberspace and coordinates operational responses to cyber events of national importance. Defense maintains the Network Operations Centre to protect and manage the security of its own networks. However, there is no publicly available strategy or policy position to guide Defense and the ADF's approach to cyber threats.
Cambodia	It seems the Cambodian Armed Forces have at least a superficial involvement with cyber policy and security, although the extent and detail of that involvement remain unclear in open-source material. Regardless of the level of defense force involvement, it is understood that Cambodia has a very limited capability to defend against cyber attacks.
China	Open-source reporting indicates that the People's Liberation Army (PLA) has several bureaus that actively conduct cyber espionage operations. The PLA has published several doctrinal information and development articles and monographs on information warfare and the role of cyber capabilities in military operations. However, there seems to be a lack of coordination of these activities within the PLA.
India	The Indian military is aware of cyber threats and has established several organs to address them, including Defense CERT, the Army Cyber Security Establishment, the Defense Information Warfare Agency, the Cyber Security Laboratory, and the Military College of Telecommunication Engineering. The establishment of a Cyber Command has been announced although it is unclear whether this has been implemented. There is a lack of stated policy direction for military cyber capabilities.
Indonesia	The Indonesian Defense Minister announced plans to establish the Cyber Defense Operations Centre to coordinate national cybersecurity efforts, including service-specific work by the Indonesian military on cybersecurity. The center is slated to draft a national doctrine on cybersecurity and conduct implementation strategies across defense and other departments. The creation of a dedicated "cyber army" has also been proposed. The force will consist of elite membership embedded in the various branches of the Indonesian military to protect domestic networks against cyber attack although it is unclear what progress has been made on this initiative.

Japan	The recent Japanese National Security Strategy outlines Japan's interests in cyberspace, including means to address current limitations in Japanese cyber capabilities. The Japan Self-Defense Force (JSDF) Command, Control, Communications and Computer Systems Command is charged with the development of national cyber defense capabilities. Under the command, the JSDF established a Cyber Defense Unit. The defense force is considered as having the necessary structures in place for cyber operations. The JSDF is working to improve its capability, especially through cooperation with the US but a shortage of qualified personnel, an inability to respond to attacks, weak capabilities, and problems in information sharing within the force remain areas of concern.
Malaysia	Reports indicate that the Malaysian Armed Forces have begun to develop capabilities to protect national assets, including from cyber threats, and the defense minister has publicly supported the development of an ASEAN master plan for cybersecurity in Southeast Asia. However, there is a lack of clear policy direction for the development of cyber capabilities.
Myanmar	The Defense Services Computer Directorate, under the Army Chief of Staff, encompasses network-centric warfare, military-oriented cyber capabilities, and electronic warfare. The army's military strategy has been expanded to include cyberwarfare as part of "people's war under modern conditions." Military Affairs Security (formerly the Directorate of Defense Services Intelligence) also possesses a cyber unit, but is more politically focused, carrying out monitoring both domestically and internationally. There are suggestions that the unit's capability has grown exponentially in recent years with the assistance of other countries in the region. Russia and China have provided training to officers, and Singapore and China have both provided physical infrastructure support.
North Korea	The North Korean military is believed to have highly developed cyber capabilities and a well-organized and extensive education and research program to support future operations. It is believed that Unit 121 is its primary offensive cyber force, that its military has successfully infiltrated South Korean government and private sector systems, and personnel estimates range from 300 to 3,000 people. There is little understanding of the military's defensive capabilities.
Papua New Guinea	Despite recent attempts to bolster the strength of the Defense Force, which has limited capabilities and resources, cyber issues have traditionally not been a priority. The 2013 Defense White Paper mentions establishing a defensive "Cyber Cell" to protect a yet to be developed "Integrated ICT Network," but outlines no timelines or implementation strategies. Clear evidence of military cyber policy and capacity in cyber operations remains limited.

Philippines	The Armed Forces of the Philippines have created a Security Operation Center with a primarily defensive role, protecting military systems. However, the extent of implementation remains unclear.
Singapore	The Singaporean Armed Forces have established a Cyber Defense Operations to protect domestic military networks. This indicates an awareness of cyber risks and that work is underway to address them. However, there is no publicly available Armed Forces strategy or policy on how the armed forces will engage with cyber threats.
South Korea	South Korea has a capable military cyber capacity. The Defense Information Warfare Response Center of the Defense Security Command protects military networks, while the Cyber Command unit handles wider online security. South Korea has both defensive and offensive capabilities and in February 2014 announced its intention to develop offensive cyber capabilities specifically to target North Korea's nuclear program. However, there are recent allegations of military cyber unit interference in national elections, and a new Cyber Defense Department (due to launch in May 2014) aims to stop these domestic interference issues. The new command will be under the Joint Chiefs of Staff with responsibility for all cyber warfare missions. It will include an oversight committee and a whistleblower program.
Thailand	The Thai military has limited capability and authority on cyber issues but its leadership has expressed an interest in developing legislation for the operation of a cyber army. Thailand hosted the 2013 USPACOM Cyber Endeavour program, which focused on communications and IT interoperability.
United States	The DoD role in cyberspace is largely concerned with signals intelligence, the defense of.mil domains, and offensive and defensive military cyber operations. Cybersecurity has been identified as a national security priority in the National Security Strategy, and DoD has published a Strategy for Operating in Cyberspace to guide its cyber efforts. The military possesses sophisticated offensive, defensive, and surveillance capabilities but internal coordination and governing policies concerning those operations could use further development.

Source: Adapted from material within the Australian Strategic Policy Institute's report, Cyber Maturity in the Asia-Pacific Region, April 2014.

Notes

1. Richard A. Bitzinger, "Outline Document, MacArthur Foundation-supported Workshop," presented to the conference on "The Potential Military Impact of Emerging Technologies in the Asia-Pacific," 8 January 2013, Singapore.
2. Michael N. Schmitt, "Tallinn Manual on International Law Applicable to Cyber Warfare," March 2013.
3. Ralph Langner, Cybersecurity Consultant, Presentation, 28 September 2012, http://www.youtube.com/watch?v= v1EcziU_AtY last accessed 23 October 2015.
4. Langner, 28 September 2012.
5. Langner, 28 September 2012.
6. Raiu, Cyber Terrorism—Industry Outlook; Langner, 28 September 2012.
7. European Parliament Committee on Foreign Affairs, *Draft Report on Cyber Security and Cyber Defence (2012/2096(INI)*, p. 7, 22 June 2012.
8. Meyer, Diplomatic Alternatives; See also EP Committee on Foreign Affairs, Draft Report Cyber Security.
9. Langner, 28 September 2012. "Cybered Conflict" is the turn of phrase used by Chris Demchak and Peter Dombrowski, U.S. Naval War College.
10. James Clapper, "Worldwide Threat Assessment," 2013.
11. Clapper, "Worldwide Threat Assessment."
12. Schmitt, "Tallinn Manual."
13. Schmitt, "Tallinn Manual." Jamming is known as saturating a receiver with noise or false information.
14. Schmitt, "Tallinn Manual."
15. Schmitt, "Tallinn Manual."
16. Langner, 28 September 2012.
17. Langner, 28 September 2012. See also Alexander Klimburg, Austrian Institute of International Affairs, "Mobilising Cyber Power," *Survival*, Vol. 53, No.1, February–March 2011, p. 41.
18. Langner, 28 September 2012.
19. United Kingdom Ministry of Defence, *Strategic Trends Programme, Regional Survey—South Asia out to 2040*, October 2012.
20. James Lewis, CSIS, "Hidden Arena."
21. Pascal Vennesson, "Dimensions of War and Strategy," 15th Asia Pacific Programme for Senior Military Officers—The Future of War, RSIS Singapore, 5 August 2013.
22. Schmitt, "Tallinn Manual."
23. Schmitt, "Tallinn Manual."
24. UK MoD, South Asia out to 2040.
25. UK MoD, South Asia out to 2040.
26. UK MoD, South Asia out to 2040.
27. American Forces Press Service, "NSA Chief: Cyber World Presents Opportunities, Challenges," 10 July 2012, http://www.defense.gov/News/NewsArticle.aspx?ID= 117060, last accessed 23 October 2015.
28. Adam Segal, "Huawei, Cyber Security, and U.S. Foreign Policy," 10 October 2012, http://blogs.cfr.org/asia/2012/10/10/huawei-cybersecurity-and-u-s-foreign-policy/, last accessed 23 October 2015.

29. Ahmed Hashim, "Warfare in New Domains: The Future of Asymmetric Operations and Information Warfare," 15th Asia Pacific Programme for Senior Military Officers—The Future of War, RSIS Singapore, 5 August 2013.
30. Hashim, "Warfare in New Domains."
31. McAfee, "2013 Threats Predictions Report." See also James Clapper, "Worldwide Threat Assessment," 2013.
32. Australian Strategic Policy Institute, International Cyber Policy Centre, "Cyber Maturity in the Asia-Pacific Region 2014," April 2014.
33. ASPI, Cyber Maturity Asia-Pacific.
34. Han Nack Hoon, "South Korea's Defence Reform Plan: A Case of Bad Timing?," RSIS Commentary, 12 October 2012.
35. Adam Segal, "U.S. and China in Cyberspace: Uneasy Next Steps," 18 June 2012, http://blogs.cfr.org/asia/2012/06/18/u-s-and-china-in-cyberspace-une asy-next-steps/, last accessed 23 October 2015.
36. Richard Bitzinger, "The Challenge of Strategic Ambiguity in Asia," RSIS Commentary, 13 March 2014.
37. Michele Flournoy, Former U.S. Under Secretary of Defense for Policy, "U.S. National Security Priorities in President Obama's Second Term," RSIS Singapore, 5 December 2012.
38. Paul Meyer, Former Ambassador to the UN, "Diplomatic Alternatives to Cyber-Warfare: A Near-Term Agenda," *RUSI Journal*, February/March 2012 Vol. 157, No. 1, p. 16.
39. Schmitt, "Tallinn Manual."
40. UK MoD, South Asia out to 2040.
41. Segal, Huawei, Cyber Security, and U.S. Foreign Policy.
42. Charles Hutzler, "China Defense Ministry Denies U.S. Hacking Charges," Wall Street Journal, 20 May 2014.
43. Franz-Stefan Gady, EastWest Institute, "The 'Cyber Weapons Gap': What do we really know about China's cyber warfare capabilities?," 21 December 2012, http://www.chinausfocus.com/political-social-development/the-cyber -weapons-gap-what-do-we-really-know-about-chinas-cyber-warfare-capabili ties/, last accessed 23 October 2015.
44. Institute for Security Technology Studies, Cyber Warfare: Analysis of Means.
45. Gady, "Cyber Weapons Gap."
46. Command, Control, Communications, Computers, Intelligence, Surveillance and Reconnaissance.
47. Gady, "Cyber Weapons Gap."
48. Alexander Klimburg, Austrian Institute of International Affairs, "Mobilising Cyber Power," *Survival*, Vol. 53, No. 1, February–March 2011, p. 48.
49. Gady, "Cyber Weapons Gap." See also Institute for Security Technology Studies, Cyber Warfare: Analysis Means, and Klimburg, Mobilising Cyber Power.
50. Bitzinger, "Challenge of Strategic Ambiguity in Asia."
51. Adam Segal, "China, International Law, and Cyberspace," 2 October 2012, http://blogs.cfr.org/asia/2012/10/02/china-international-law-and-cyberspace /. See also Meyer, Diplomatic Alternatives, last accessed 23 October 2015.
52. EP Committee on Foreign Affairs, *Draft Report Cyber Security*.
53. United Nations (A/66/359), *Letter dated 12 September 2012 from the Permanent Representatives of China, the Russian Federation, Tajikistan and Uzbekistan to the*

United Nations, 14 September 2012. See also Segal, "China, International Law, and Cyberspace."

54. Meyer, Diplomatic Alternatives (reference to Ambassador Wang Qun's speech, First Committee of the 66th Session of UN GA, 19 October 2011).
55. UK MoD, South Asia out to 2040.
56. UK MoD, South Asia out to 2040.
57. Clapper, "Worldwide Threat Assessment."
58. UK MoD, South Asia out to 2040.

10
The (Over)Promise of Modern Technology

Bernard Loo

Most of us are familiar with the argument about technology as a force multiplier, allowing states with limited strategic resources—manpower, strategic depth, and so on—to potentially overcome these *potential* strategic shortfalls. The reason for the emphasis on "potential" is deliberate: the technology-as-force-multiplier argument is focused on the potentials for most states. For Singapore, definitely—Singapore as a modern independent state has never had to taste war; the Singapore Armed Forces (SAF) has never had to be tested. For the SAF, therefore, their argument about technology helping the organization fulfill its mission of protecting the territorial integrity and sovereignty of Singapore is a potential argument. Unless and until war comes to Singapore, unless and until the SAF has to actually carry out its mission statement of defeating potential aggressors swiftly and decisively, we will never know if the SAF can actually do what it sets out, what it claims to do.

However, the argument in this chapter starts from the viewpoint that the qualitative advantages any armed forces should seek might actually best come not from technology, but from training. In other words, maybe the best and most lasting force multiplier effect comes from quality training, not from cutting-edge technology.

Here's the fundamental problem: technology is *necessarily short-lived*. Does anyone use floppy disks or 486-chip computers? Who still walks around town listening to music from a portable CD player? Yet, a decade or so ago, these were still fairly cutting-edge technologies. The difference is this—technological change is accelerating. Anyone who has read Martin van Creveld's *Technology and War* knows this. Military technology once moved at a pace that would have made tortoises, turtles and snails seem Speedy Gonzales-like. Come the 19th Century, however, the pace of change in military technologies began to accelerate, and as the 20th Century wore on, that accelerating pace itself accelerated even faster.

What this therefore means is that for any armed forces seeking to maintain a technological advantage over its putative adversaries, accelerating technological changes means that any extant technological advantage (already naturally temporary) is increasingly short-lived. That armed forces is going to spend an increasing amount of time, money, and other resources looking for the next technological advantage, and the next, and so on.

Furthermore, new technologies, particularly in the military domain, are increasingly expensive. The F-35 Joint Strike Fighter was initially envisaged as a low-cost and ubiquitous air combat platform. Well, that vision pretty much no longer exists, certainly the part about being low cost. And as advanced economies slow down in terms of their annual growth rates (that is, assuming no Lehman Brothers or Greece-style meltdowns), it means that the absolute amount of money that can be dedicated to military expenditure is going to slow down, in terms of growth rates as well. It may be difficult to prove, but it is arguable that the costs of emerging military technologies are out-stripping economic growth rates. Which means armed forces will be able to afford fewer and fewer of these new technologies—something that some have called "structural disarmament." Indeed, new technologies may be more capable than the ones they replace, but fewer platforms also means that the loss of a single platform represents a greater percentage loss of potential combat power.

Moreover, as the preceding paragraphs suggest, the putative adversary right now might be at a technological disadvantage, but the pace of technological change means that that disadvantage is going to be increasingly short-lived. If this is correct, therefore, it seems that technology is not going to help us out of a strategic conundrum. Human qualities are probably going to be the more lasting solution.

The key issue that we should address next is the follow-on from the increasingly temporary nature of technological advantages. For armed forces that rely on technological advantages to help them gain a force multiplier effect, to help them overcome other inherent strategic disadvantages—whether small population, small size, whatever—this reliance on technological advantages is going to mean an ever-increasing pace of technological change in their military hardware. Once, investment in a new weapons technology might have accrued force multiplier effects for maybe two or three decades; now, that force multiplier effect is likely to last significantly less. It means an almost-constant search for the next technological development.

At the same time, because military technologies are getting increasingly expensive (again, the so-called structural disarmament argument), the numbers of new combat platforms that can be bought is necessarily being reduced. For example, the Republic of Singapore Air Force's F-15SG does not just cost more than the F-5 aircraft it replaced, it cost exponentially more. As a consequence, the RSAF could not replace F-5 with F-15 on a one-for-one basis: in other words, if the RSAF wanted to retire a hypothetical 100 F-5s, it could not buy 100 F-15s as the replacement; and, in fact, the number of F-15s bought (24) was significantly less than the 100 F-5s these aircraft replaced.

Now, people may argue that F-15s are significantly more capable aircraft than F-5s. Certainly, one F-15 probably does have the combat power of several equivalent F-5s. However, it is also true is that if, for whatever reason, one F-15 is lost, that loss is much more significant (in terms of percentage), than the loss of one F-5.

In other words, it is very difficult to measure the effectiveness (more accurately, the *potential* effectiveness) of combat systems. In this regard, two viewpoints are particularly germane. The first comes from Matthew Schofield, in his 2012 article on the US Navy.[1] As Schofield notes, the US Navy presently has a total of 286 ships, down from its Cold War high of 594 ships in 1987. At the same time, however, he notes that it is important to remember that right now, the United States does not face a peer competitor. In this regard, Schofield supplies an insightful quote by US Naval War College Associate Professor James Holmes: "We judge naval combat power on a relative scale ... That's why 'the Navy is smaller than it has been since 1917' and 'the Navy is bigger than the next 13 navies combined' both contain a grain of truth but are basically factoids. Numbers count; the tonnage of ships counts; but these one-liners tell us little."[2]

The second viewpoint comes from James Holmes' article about the Chinese navy.[3] In this article, Holmes notes (as his number three priority), that the Chinese navy ought emphasize the so-called "unsexy ships," by which he means a fleet logistics capability that would allow Chinese naval vessels to remain at sea indefinitely, allowing them to surmount "the tyranny of distance." But what is really instructive is Holmes' first two priority items for the Chinese navy: (1) go to sea a lot, and (2) act like a blue-water navy. As Holmes notes, his most important two priorities are what he terms as the "human factor in seafaring and maritime combat" (and both, by the way, are largely non-quantifiable elements).

Military strength, strategic effectiveness, combat power: in a sense, all three concepts are the same. Military hardware is undeniably important, *but it is not the only determinant*! Qualitative variables cannot and must not be omitted from the "equations" we use to "calculate" these concepts. One is reminded once again of the Gurkha soldier, Dipprasad Pun, who single-handedly turned back a team of over 30 Taliban fighters assaulting his position. One is also reminded of how, in 1991, most strategic analysts thought defeating Saddam Hussein was not going to come easy, based purely on what military hardware the Iraqi military could deploy—and how wrong they all were!

Finally, we need to focus on how military organizations can begin to approach this entire question of accelerating technological change, and its implications for force structure decision-making. I think the issue can be set up as a straightforward contest between the "best" versus the "good enough." One of the best analogies of this argument is the contrasting philosophies of tank design between Soviet Russia and Nazi Germany during the Second World War. Few people will dispute that Nazi Germany produced probably the most technologically sophisticated tanks during the Second World War. As a consequence of the technological sophistication, Nazi Germany had to sacrifice quantity for technological quality. In contrast, Soviet Russia produced the T-34, which although not as technologically advanced as their German counterparts, could be produced in very large numbers; the Russians sacrificed quality for quantity, in other words. At the greatest tank battle of that war, at Kursk, quantity prevailed over quality. But then again, to paraphrase Napoleon, enough quantity has a quality of its own.

When it comes to decisions about force structures, military planners always face this question: do I want the "best" (of which one can perhaps only afford, say, 10 platforms), or do I go for the "good enough" (of which I can afford, say, 40 platforms)? Given that technological change is accelerating, it means that the shelf life of any given technology is shorter than its predecessors. We ought to remember that the acquisition process itself incurs costs: one is not simply paying for a new platform, how one acquires that platform also has to be paid for. The shrinking shelf life of new technologies allied to the perceived need to remain at the technological cutting edge combine to create a potential financial storm. What makes it worse is that the platforms one just acquired are going to give one a technological advantage that lasts for a very short time.

Maybe, therefore, the solution for military organizations is to go for the "good enough." And to blend that "good enough" technology with the best training you can possibly give your personnel. Looking at vignettes that come from Coalition operations in Afghanistan especially, what appears to have saved thousands of Coalition soldiers was not their superior technology against their Taliban counterparts; rather, it was their superior training that mattered the most. Ultimately, a "good enough" technology in the hands of a superbly trained soldier generates a combat effect far greater than a "best" technology in the hands of a soldier who doesn't know what to do with it. Again, we saw this in the 1991 Gulf war; most military analysts initially thought Iraq was not going to be an opponent that rolled over easily; but then the shooting started, and the lack of quality in the Iraqi soldiers became evident. In conclusion, then, maybe, just maybe, that could be the way forward for small armed forces with very limited resources in manpower and finances.

Notes

1. Matthew Schofield, "To Tally the Navy's Strength Requires More than Math," *McClatchy Newspapers*, 23 October 2012 (http://www.mcclatchydc.com/2012/10/23/172369/to-tally-the-navys-strength-requires.html#storylink=cpy, accessed 14 May 2014).
2. IbidSchofield, "To Tally the Navy's Strength Requires More than Math".
3. James R. Holmes, "Top 5 Things China's Navy Needs…to Be a Blue-Water Navy," *The Diplomat*, 5 August 2012 (http://thediplomat.com/2012/08/top-5-things-chinas-navy-needs-to-be-a-blue-water-navy/, accessed 14 May 2014).

11

The Future Is upon Us: Failed Predictions, Boiling Frogs, and Gun Printers

Paul T. Mitchell

Militaries are future-oriented institutions.[1] They consider the impact of advancing technology, changing political dynamics, and the implications of global economic shifts. Yet, as Niels Bohr reminds us, predictions are difficult, especially about the future. Very often, our predictions say more about our present: as Yogi Berra said, the future ain't what it used to be.

Presently, military forces are considering the impact of the conclusion of major military operations in South Asia, the turmoil in the Ukraine, maritime events in the South China Sea, and the growing strategic assertiveness of both China and Russia. Many of these events have called into question assumptions made in terms of how future strategic relations will unfold. In the United States, the recently announced "Asia Pivot"[2] is itself based on a reassessment of geostrategic priorities, driving military strategies like the Joint Operational Access Concept[3] and the so-called Air Sea Battle.[4] At the Canadian Forces College, the Majors on the Joint Command and Staff Program conduct an annual exercise called Global Powers, which requires them to examine the shifts in power between the major states and emerging ones like the "BRICS" in an effort to understand the environment in which the Canadian Armed Forces may be called to operate. The estimates that students produce are often remarkably conservative in nature, few of which call for any major transformations in global relations, which are defined in terms of other powers replacing the role the US plays currently. In essence, this is not at all surprising. The track record even of experts in predicting the future is decidedly poor.[5] Despite decades of intensive regional study, events like the collapse of the Soviet Union and 9/11 came as total surprises, so the students share plenty of company in expecting that the future will look more or less like the present.

Nevertheless, the future is a real problem for militaries. Get it wrong and you lose a war, or worse. Historically, this has not always been so. During ancient times, the advance of technology was remarkably slow. Agricultural societies practiced a form of warfare that did not change over periods of millennia. The equipment of Egyptian forces during the Middle Kingdom did not dramatically differ from that of Alexander the Great or even later Roman legions. The technology employed by Dark Ages Vikings did not dramatically differ from that used by Henry the fifth at Agincourt. The combination of light infantry, heavy infantry, light cavalry, and heavy cavalry were present in varying amounts in all armies such that tactical affairs were essentially a game of rock-paper-scissors, and strategy was essentially about looking at what had been done in the past to figure out what should be done in the future.

Industrialism changed all of that, however. The introduction of firearms began a period of technological development, fueled by the growth of both capital markets and the concentration of power in states. As the historian Charles Tilley remarked of this dramatic shift: "War made the State and the State made War."[6] Scientific methods of management, anticipating the ergonomic studies of Frederick Taylor, and the industrial production line emerged to guide the employment of the new military formations and the science of geometry was employed to create new fortifications taking advantage of firearms and even the movement of forces to and from the battlefield. Science, rather than history, became the feature animating military power. Militaries could get caught by surprise with striking results if they did not pay attention to how technological and political developments were affecting the conduct of war. Perhaps the best recent example of this was during the Battle of France when French and British armies were surprised by the different ways in which the Wehrmacht employed armor, aircraft, and wireless radio technologies.

But if a new information age is upon us, what effect could such a social change have upon the conduct of warfare? During the 1990s, following the spectacular feat of arms in the liberation of Kuwait, speculation began to surface about a so-called "Revolution in Military Affairs"[7] and the role played by emerging information technologies.[8] Global Positioning Systems, Missile Defence, Precision Guided Munitions, battlespace management systems aboard AWACS and JSTARS radar planes, and stealth technology affected the war in ways that suggested that a major change, perhaps not unlike the German introduction of blitzkrieg tactics. The outcome of this debate was a vision of futuristic air traffic control-like technologies applied to the battlefield,[9] where everything could be seen, tracked, targeted and destroyed. It suggested a level of god-like control over battlefield events and led to operational concepts like "information dominance" and "decision superiority."[10] The experiences of both Iraq and Afghanistan revealed that no matter how efficient information technology could make battlefield operations, it had very little effect on strategic outcomes. War remained a messy, brutal affair as it always has been. The future, once again, failed to conform to what we expected.

The trouble with most attempts to predict the future is that it is essentially unknowable. What we call the "future" is actually the contingent outcome of a multiplicity of events, decisions and actions by innumerable actors. Looking back, there seems to be a story, what we call history, which suggests a certain inevitability to how events ultimately turn out. But this is a logical fallacy: looking at diaries, journals, and interviews of those in the past, there is always a sense that people have little idea of the ultimate significance of the events taking place all around them.[11] Thus, most predictions are often extrapolations of what presently exists. This sort of methodology leads to estimates of flying cars and space colonies, and fails to detect the Arab Spring.[12]

In the 1950s, The US Army was struggling to deal with the imagined effects of nuclear weapons on the battlefield. As Andrew Bacevich writes in his minor classic, *The Pentomic Era*, "New technology, changing views of the nature of war, and the fiscal principles of the Eisenhower administration produced widespread doubts about the utility of traditional land forces."[13] In other words, it was a time not unlike the present. The US Army's "Pentomic Division" concept responded to these contextual pressures by imagining a future force that could survive the conditions of what an atomic battlefield was thought to be like.

> The importance of dispersing to improve survivability against nuclear weapons exercised the greatest influence on the structure of this new organization. Dispersion meant that units within a division necessarily would fight with greater autonomy than they would have in earlier wars. On the deep and fluid battlefield that Army theorists envisioned units would find themselves on their own – seldom tied in with friendly units on their flanks, unable to count on higher echelons to assist with either direction or materiel. Such circumstances would require maneuver units that were self-contained and self-sustaining.[14]

Like cells in a body, the loss of any one would not preclude the remainder from carrying on. The new organization would be built upon a new concept, the battlegroup, a formation of roughly battalion size, made up of five companies, rather than the traditional three. Such a wealth of resources would enable it to fight in a "non-linear" battlefield where ideas of front and rear would no longer make any sense.

The Pentomic Division did not survive the Kennedy administration, with its focus on limited wars and "flexible response." Massive retaliation, the strategy on which the Pentomic Division was based, became increasingly incredible as the Soviet Union developed its own nuclear forces, and European allies were hardly enamored with American plans to fight a nuclear war on their soil. Still, the notion of high-tech army units fighting on non-linear battlefields is an idea that the US seemingly finds irresistible. In the late 1990s, the army developed the concept of the Future Combat System (FCS). Just as nuclear weapons provided the motive force in the 1950s, in the 1990s, the army sought to grapple with the changes that information technology was bringing to the battlefield.

The army's problem, as it saw it at the beginning of the new millennium, was mobility. While the Air Force and the Navy were inherently expeditionary in nature, the large, armor-heavy units that made up America's land forces were slow and difficult to deploy. Information technology would produce superior knowledge of the battlespace, however, enabling land forces to achieve dramatic effects in terms of their ability to both position themselves and engage enemy forces. Multidimensional assaults from land, sea, and air taking place simultaneously throughout the entire volume of the battlespace would not be easily countered by opposing forces, creating paralyzing effects in the minds of enemy decision makers.[15] The growing lethality of long range precision weapons would continue the expansion of the battlespace by enabling the dispersion of scarce military resources over larger and larger areas throughout it. Precise knowledge of the friendly and enemy forces would permit units to customize themselves to specific operational demands and reduce the need to "fill space with forces and direct fire weapons." As FM-1 put it, "The goal of future Army operations will be to simultaneously attack critical targets throughout the area of operations by rapid maneuver and precise fires to break the adversaries will and compel him to surrender."[16] A so-called "Quality of Firsts" would be generated by these relationships: information-age land forces would be able to "see first, understand first, act first, and finish decisively." For the first time in the history of warfare, such enabled land forces would be able to consistently pick and choose the time and place of battle, eliminating dangerous and unpredictable "meeting engagements."[17]

However, such information-age land forces would be qualitatively different from those of the past. The information systems that would make them powerful offensive forces would come at the expense of armor and to a certain extent, self-reliance: these units will be profoundly reliant on the coordination of services from other, typically more distant supporting units. While information would alleviate the dichotomies raised between light and heavy forces, it would not eliminate them altogether. Lacking sufficient armor to independently defend themselves, the new units would need to draw on the capabilities of other forces, possibly in different services, to provide complete force protection. Information distributed through networks would heighten the offensive power of land forces, but would in turn become absolutely critical for their defense as well, in the key role of coordinating the efforts of dispersed and individually weak forces.[18]

In order to realize its vision for future warfare, the army pursued a number of programs, most visibly in terms of the FCS, Stryker Brigades, and Landwarrior (an "augmented reality" infantry system). Based around a common light armored chassis from the MOWAG Piranha family of vehicles, the Stryker Brigades were to rely heavily on distributed sensors to provide superior situational awareness and long range precision weapons for immediate striking power, compensating for the vehicles' relative lack of armor.[19] While lacking in armor, it was hoped that the brigade would bring the same capability as a scaled-down division.[20] Like its Pentomic predecessor, the FCS failed to survive the end of its sponsoring administration, being cancelled by the incoming Obama administration. However, many of its programs had already been heavily criticized due to their cost and developmental problems, and it had long been on life support when it was finally terminated.[21]

The latest technological revolution promising to fundamentally change the nature of warfare is in the field of robotics. The landing of Boeing's X-47B UAV on the deck of the USS George H. Bush in the summer of 2013 has provoked considerable interest in autonomous weapon systems.[22] Feats such as these have sparked the imagination of many and raise questions about the progress of autonomy in weapons. After the landing, the blogosphere came alight with speculation about the development of autonomous robots on the battlefield.[23] David Betz from the King's College War Studies department suggested that technology, policy, and military practice are all leading towards a future of autonomous robots in the conduct of warfare. Steve Metz of the US Army War College drew similar conclusions, noting the growing challenge of recruiting (and affording) sufficient numbers of troops to deal with the many challenges confronting the United States.[24] Finally, LSE professor, Christopher Coker's *Waging War Without Warriors* discusses so-called "Transhuman Warfare" in terms reminiscent of the *Terminator* movie franchise.[25]

We have to be clear—there is an enormous difference between today's Predator B drone aircraft and autonomous robots. UAVs, while capable of taking off and landing by themselves, and flying unassisted to patrol areas, are not true autonomous robots. In combat, they are under the full control of humans and they can neither identify a target nor launch a weapon on their own. Indeed, while the issue of targeted killings by drones is worthy of debate,[26] there is no functional difference between an airstrike conducted by a manned F-16 and that of a Predator B: a human being pulls the trigger in both instances. It just so happens that with a UAV, that human being is located thousands of miles away, rather than being on the scene. In both cases, the result is the same.[27]

Further, we should acknowledge that modern militaries have a legitimate interest in robotics. The economic and social costs of warfare are spiraling ever greater and while we may decry the use of force, our governments continue to see great utility in employing it for a growing variety of purposes.[28] The cost of soldiering is increasingly expensive, however. Recent reports estimate that it costs between US$850,000 and US$1.4 million per soldier per year to support Afghan operations and the separate bill for training and social benefits of each soldier are equally as large.[29] Rising costs have basically led to smaller forces. As Metz argues, robots may be a way of dealing with this problem (As an aside, John Ellis' classic, *The Social History of the Machine Gun* makes a similar argument in terms of the introduction of automatic weapons in the late 19th century—higher rates of fire enabled smaller forces to take on more numerous enemies in colonial conflicts).[30]

As with the case of both the Pentomic Division and the FCS, the present interest in autonomous weapon systems is being couched in terms of operational necessity. In their report "20YY: Preparing for War in the Robotic Age," Robert Work and Shawn Brimley make similar types of arguments in terms of technological progress and its effects on the conduct of war that were made under the concepts of "nuclear age" and "information age." While they disavow notions of perfect battlespace awareness and reaffirm the "human" aspects of war, the future they envisage is still fundamentally directed by technological developments.

"To allow the U.S. military both to weather these buffeting winds of change and to capitalize on real opportunities to extend America's technological edge, DOD must urgently spur new thinking and research on the changing nature of warfare and the types of new systems, organizations, and operational concepts needed to conduct it."[31] It is hard not to be cynical that plans for robot equipped armies will be any more successful than those designed to fight on nuclear and information battlefields.

As opposed to these imagined futures, real revolutions either emerge from weak signals approaching from the horizon, like that of a disenfranchised fruit seller immolating himself, or from the slow development of material forces wherein incremental changes ultimately result in entirely new circumstances.[32] The growth of information technologies *is actually* an appropriate example of this. Writers as diverse as Marshall McLuhan, Peter Drucker, Alvin Toffler, Manuel Castells, and Daniel Bell, writing across several decades, have all called attention to the effect of information technology on the emergence of new material and social conditions that will transcend the old structures of the scientific and industrial revolutions.[33] In fact, as Vincent Moscoe points out, the transformative promises of information technology date back as far as the invention of the telegraph, in terms that are similar to today's predictions surrounding the Internet.[34] Indeed, some have even argued that social-media-like effects date as far back as the Gutenberg press.[35] Like the proverbial frog in the pot of water, we don't realize change is upon us until it is too late.

In the agricultural age, according to Castells, mankind was simply surviving as best as possible the harsh conditions of the natural world. Tyrannical imperial political systems reflected these realities, organized around maintaining control of populations forever on the fine edge of disaster. Warfare sought to increase the resources to sustain subject population (and increase the labor available to work) or to eliminate the excess through the organized slaughter of war. In the industrial age, mankind learned to "conquer" nature and biological energy was replaced by mechanical and chemical varieties.[36] Science mobilized and regularized the power of human creativity and industry, and capitalism developed a growing variety of products in a continuous and reflexive process of development.[37] One thing led to another and, in turn, built upon what went before.[38] Knowledge became rationalized and ordered into a growing set of categories, which in turn, demanded the growth of specialization. The state also became increasingly ordered and extended that order outwards to society itself.

Castells concludes that the information age will be a social age.[39] However, such a development may not be as progressive as his observation seems. The spread of ideas and our growing ability to control and manipulate information may lead to the intensification of differences amongst us, rather than their diminishment.[40] Recommendation services in systems like online book merchants and music systems may simply reinforce our pre-existing wants rather than allowing us to discover new ideas; social media tends to categorize us within the ideas with which we are already comfortable, echo chambers in many cases. The explosion of conspiracy theories and scandals, the cynicism surrounding political, religious, and professional authority all suggest a growing skepticism towards any claims of truth. These social divisions and the growing difficulty in arriving at political compromises necessary for social cohesion evident not only in recent US elections but even in Canadian political debates all point to the divisive effect of growing access to information. One way in which the coming age may be more social is in the growth of contested knowledge and the social process of trying to establish both "truth" and political order.

Many have argued that information technology is inherently liberating and will lead to democratic outcomes. However, if information is corrosive to both knowledge and collective agreement over truth, it may be more the source of conflict than democracy. While information technology may not necessarily lead to liberal outcomes, it is true that it generates agency for individuals, institutions, and perhaps even technology itself.[41] 3D printers, also a form of information technology, are another modality in which agency can be liberated. If you can imagine it, you can build it, from Lego bricks to buildings.[42] This can liberate the creation of products from industrialized processes, which seems to promise flights of inventive fancy subject only to the limitations of imagination. But what if you want to build assault guns, rocket launchers, and high-grade explosives? What if you can build your own arsenal and equip your own army of "followers"?[43]

If the industrialized age was one of rationalized control and ordered process into which the individual had to conform, the information age may be one of socially designed movements to fit your own specific lifestyle fashion and those who agree with you (and to hell with the rest). Rather than the "levée en masse," we may have the levée sélective. That would be a real revolution in military affairs, staring at us in the face, but completely unobserved, engrossed as we are with the latest in tablet computing.

Notes

1. The views expressed here are those of the author alone and do not represent those of the Canadian Forces College or the Department of National Defense.

2. Department of Defense, *Sustaining US Global Leadership: Priorities for the 21st Century Defense* (Washington DC: US GPO, 2012), http://www.defense.gov/news/Defense_Strategic_Guidance.pdf (accessed 29 May 2014).

3. Department of Defense, *Joint Operational Access Concept* (Washington DC: US GPO, 2012), http://www.defense.gov/pubs/pdfs/JOAC_Jan%202012_Signed.pdf (accessed 29 May 2014).

4. Michael Raska, "The Air Sea Battle Debate: Operational Consequences, Allied Concerns," *Defense News*, 30 October 2012, http://www.defensenews.com/article/20121030/DEFFEAT05/310300008/Air-Sea-Battle-Debate (accessed 29 May 2014).

5. Phillip Tetlock, *Expert Political Judgement: How Good Is It? How Can We Know?* (Princeton: Princeton University Press, 2005).

6. Charles Tilly, *The Formation of National States in Western Europe* (Princeton: Princeton University Press, 1975).

7. See http://www.comw.org/rma/fulltext/overview.html (accessed 29 May 2014) for a comprehensive overview of literature associated with the RMA.

8. Alan D. Campen, *The First Information War: The Story of Communications, Computers and Intelligence Systems in the Persian Gulf War* (Fairfax VA: AFCEA Press, 1992).

9. Vice Admiral Arthur K. Cebrowski and John Gartska, "Network Centric Warfare: Its Origin and its Future," *Proceedings*, Vol. 124, No. 1, January 1998.

10. Jim Winters and John Giffen, "Information Dominance vs. Information Superiority," http://www.iwar.org.uk/iwar/resources/info-dominance/issue-paper.htm (accessed 29 May 2014).

11. Rob Dreher, "The Narrative Fallacy," *The American Conservative*, 7 August 2012, http://www.theamericanconservative.com/dreher/the-narrative-fallacy/ (accessed 29 May 2014).

12. Stuart Fox, "Dude, Where's my Flying Car (and Jet Pack and Robot Armies)?," *Popular Science*, 1 December 2008, http://www.popsci.com/scitech/gallery/2008-12/dude-wheres-my-flying-car-and-jetpack-and-fusion-energy (accessed 29 May 2014).

13. Andrew Bacevich, *The Pentomic Era* (Washington DC: National Defense University Press, 1986), p. 3.

14. Bacevich, *The Pentomic Era*, p. 104.

15. Douglas MacGregor, *Breaking the Phalanx* (London: Praeger, 1997), p. 37.

16. Lt. Col. H.R. McMaster, "Crack in the Foundation: Defense Transformation and the Underlying Assumption of Dominant Knowledge in Future War," *CSL Student Issue Paper*, Vol. S03-03, November 2003, p. 56.

17. McMaster, "Crack in the Foundation," p. 57.

18. MacGregor, *Breaking the Phalanx*, pp. 50–52.

19. McMaster, "Crack in the Foundation," p. 54.

20. Thomas K. Adams, *The Army After Next: The First Post Industrial Army* (Westport CT: Praeger, 2006), p. 184.

21. Government Accounting Office, *Defense Acquisitions: Improved Business Case is Needed for Future Combat Systems Successful Outcome* (Washington

DC: US GPO, 2006). Department of Defense, "Future Combat System (FCS) Program Transitions to Army Brigade Combat Team Modernization," News Release, 23 June 2009, http://www.defense.gov/releases/release.aspx?releaseid= 12763 (accessed 29 May 2014).

22. Brandon Vinson, "X-47B Makes First Arrested Landing at Sea," News Release, 10 July 2013, http://www.navy.mil/submit/display.asp?story_id= 75298 (accessed 29 May 2014).

23. See Christopher Newman, "'Moralization' of Technologies—Military Drones: A Case Study," *E-International Relations Students*, 2 May 2012, http://www.e-ir.info/2012/05/02/moralization-of-technologies-military-drones-a-case-study/; J. Michael Cole. "When Drones Decide to Kill on their Own," *The Diplomat*, 1 October 2012, http://thediplomat.com/2012/10/why-killing-should-remain-a-human-enterprise/; "The Next Wave in U.S. Robotic War: Drones on Their Own," *Defense News*, 28 September 2012, http://www.defensenews.com/article/20120928/DEFREG02/309280004/The-Next-Wave-U-S-Robotic-War-Drones-Their-Own?odyssey=tab|topnews|text|FRONTPAGE; Sharon Weinberger, "Next generation military robots have minds of their own," *BBC Future*, 28 September 2012, http://www.bbc.com/future/story/20120928-battle-bots-think-for-themselves (all accessed 28 May 2014).

24. Steve Metz, "Strategic Horizons: The Future of Roboticized Warfare," *World Politics Review*, 26 September 2012, http://www.worldpoliticsreview.com/articles/12372/strategic-horizons-the-future-of-roboticized-warfare (accessed 29 May 2014).

25. Christopher Coker, *Waging War Without Warriors* (Boulder CO: Lynne Rienner, 2002).

26. International Human Rights and Conflict Resolution Clinic Stanford Law School, Global Justice Clinic NYU School of Law. *Living Under Drones Death, Injury, and Trauma to Civilians From US Drone Practices in Pakistan*, September 2012, http://www.livingunderdrones.org/download-report/ (accessed 28 May 2014).

27. Elisabeth Bumiller, "A Day Job Waiting for a Kill Shot a World Away," *New York Times*, 29 July 2012, http://www.nytimes.com/2012/07/30/us/drone-pilots-waiting-for-a-kill-shot-7000-miles-away.html?pagewanted=all&_r=2& (accessed 29 May 2014). Fighter pilot Shane Riza makes a convincing argument that UAV technology, even under full human control, changes our experience of combat, making it easier to kill and thus easier to go to war. See M. Shane Riza, *Killing Without Heart* (Dulles VA: Potomac Books, 2013).

28. "Defence spending in a time of austerity," *The Economist*, 26 August 2010, http://www.economist.com/node/16886851 (accessed 28 May 2014); Edward Luttwak, "Toward Post-Heroic Warfare," *Foreign Affairs*, May/June 1995, http://www.foreignaffairs.com/articles/50977/edward-n-luttwak/toward-post-heroic-warfare (accessed 28 May 2014).

29. Larry Shaughnessy, "One soldier, one year: $850,000 and rising," *CNN Security Clearance*, 28 February 2012, http://security.blogs.cnn.com/2012/02/28/one-soldier-one-year-850000-and-rising/ (accessed 28 may 2014).

30. John Ellis, *The Social History of the Machine Gun* (Baltimore: Johns Hopkins University Press, 1975).

31. Robert O. Work and Shawn Brimley, *20YY: Preparing for War in the Robotic Age* (Washington DC: Center for a New American Security, 2013), p. 9.

32. Carole Stonebridge, "Horizon Scanning: Gathering Research Evidence to Inform Decision Making," *Conference Board of Canada*, March 2008, http://www.conferenceboard.ca/e-library/abstract.aspx?did=2491 (accessed 28 May 2014).

33. Marshal McCluhan, *The Gutenberg Galaxy* (Toronto: University of Toronto Press, 1962); Peter F. Drucker, *Landmarks of Tomorrow* (New York: Harper and Row, 1957); Alvin Toffler, *The Third Wave* (New York: Bantam Books, 1980); Manuel Castells, *The Rise of the Network Society* (Malden MA: Blackwell, 2000); Daniel Bell, *The Coming of the Post-Industrial Society* (New York: Basic Books, 1973).

34. Vincent Moscoe, *The Digital Sublime: Myth, Power and Cyberspace*, (Cambridge: MIT Press, 2004).

35. "How Luther Went Viral," *The Economist*, 15 December 2011. http://www.economist.com/node/21541719 (accessed 28 May 2014).

36. Andrew Nikiforuk, *The Energy of Slaves: Oil and the New Servitude* (Vancouver: Greystone Books, 2012).

37. Matt Ridley, "When Ideas Have Sex" The secret to human success is—and always has been—collaboration," *Design Mind*, http://designmind.frogdesign.com/articles/and-now-the-good-news/when-ideas-have-sex.html (accessed 28 May 2014).

38. W. Brian Arthur. *The Nature of Technology: What It Is and How It Evolves*, (New York: The Free Press, 2009).

39. "History is just beginning, if by history we understand the moment when, after millenniums of prehistoric battle with nature, first to survive, then to conquer it, our species has reached the level of knowledge and social organization that will allow us to live in a predominantly social world" (Castells, *The Rise of the Network Society*, 2000), pp. 508–509.

40. Nico Stehr, *A World Made of Knowledge*, http://www.google.ca/url?sa=t&rct=j&q=&esrc=s&source=web&cd=9&ved=0CF0QFjAI&url=http%3A%2F%2Fwww.crsi.mq.edu.au%2Fpublic%2Fdownload.jsp%3Fid%3D10619&ei=vgWUUI35L8T_ygGBs4HAAw&usg=AFQjCNGLE1U2Er9LPvief7WcU4D3trztCA&cad=rja (accessed 28 May 2014).

41. Michael Lewis, *The Flash Boys: A Wall Street Revolt* (New York: W.W. Norton, 2014).

42. *Future of Construction Process: 3D Concrete Printing*, http://www.youtube.com/watch?v=EfbhdZKPHro (accessed 28 May 2014).

43. Such a possibility may be already underway. Wired's Andy Greenberg discusses the growing online community committed to creating home made fire arms in the United States. Andy Greenberg, "I Made an Untraceable AR-15 'Ghost Gun' In My Office – And It Was Easy", *Wired*, 03 June 2015. http://www.wired.com/2015/06/i-made-an-untraceable-ar-15-ghost-gun/?wb48617274=F2A15BE7 (accessed 16 October 2015).

12

The Future Ain't What It Used to Be: Strategic Innovation in the Global Defense Industry

Richard A. Bitzinger

Innovation is critical, if not central, to military modernization. Throughout history, the process of innovation—that is, the process of turning ideas and invention into more effective products or services (in this case, the creation of more effective militaries)—was at the heart of gaining military superiority over a rival (or rivals). This includes the introduction of new ways of fighting (i.e., the phalanx, used to great effectiveness by the Greek city-states in antiquity), of organization (i.e., the *lévee en masse* of the French Revolution), or of technology (i.e., the so-called "gunpowder revolution" of the 16th century, or aviation and mechanization in the 20th Century). This chapter will deal mainly with last category, i.e., *technological* innovation, and its role in military modernization. Although military innovation/modernization is typically a "holistic" event, incorporating technological change with changes in organization, doctrine, and tactics, technology is still generally the starting-point for innovation, and therefore it will be central in this chapter.

At issue is whether the process of modern military-technological innovation—arguably begun around the time of the First World War (1914–1918) and which accelerated greatly during the Cold War period (1945–1990)—is beginning to fail. That is not to say that technological innovation in the defense industry is over, but rather that it has entered a new phase in which the pace of *strategic* military industrial-technological innovation—that is, dramatic and far-reaching technological change—is slowing down, or is in the midst of a "strategic pause." Consequently, we may be entering an era where "bumpy," revolutionary change (i.e., spikes in creativity and innovation) within the global defense industry, particularly at the level of research and technology (R&T), is giving way to a less radical but continuous process of innovation. If true, then this new process of innovation could have significant implications for the global arms industry, in terms of next-generation leaders and followers.

While this chapter primarily addresses the US military's recent experiences with strategic and military-technological innovation, arguments found in this chapter can be applied, to a degree, to Asian countries when it comes to the future challenges of military modernization. To a large extent, all nations seeking to gain (or retain) comparative military advantage over potential rivals will face similar challenges when it comes to harnessing militarily relevant technologies.

Disruptive vs. sustaining innovation and the revolution in military affairs

In general, there are two types of innovation: disruptive and sustaining. According to Dombrowski, Gholz, and Ross:

> *Sustaining innovations* are defined by improvements in product quality measured by familiar standards: they offer new, better ways to do what customer organization have been doing using previous generations of technology.[1]

On the other hand, *disruptive innovations* "establish a trajectory of rapid performance improvements that...overtakes the quality of the old market-leading product even when measured by traditional performance standards."[2] In other words, disruptive innovation changes nearly everything about doing business—in this case, the business of war. And nowhere is the conceptual impact of disruptive innovation on warfighting more forcefully articulated than in the theory of the "revolution in military affairs" (RMA). Above all, in the minds of its proponents and advocates, the RMA is *necessarily* a process of *discontinuous, disruptive,* and *revolutionary* change, as opposed to *incremental, sustaining,* and *evolutionary* change.[3] Andrew Krepinevich, for example, argues that an RMA occurs when

> the application of new technologies into a significant number of military systems combines with innovative operational concepts and

organizational adaptation in a way that fundamentally alters the character and conduct of a conflict. It does so by producing a dramatic increase...in the combat potential and military effectiveness of armed forces.[4]

This view of the *purposely* disruptive nature of the RMA is widely held. A 1999 RAND report, for example, defined an RMA as

a paradigm shift in the nature and conduct of military operations which either renders obsolete or irrelevant one or more core competencies in a dominant player, or creates one or more core competencies in some dimension of warfare, or both.[5]

Or, to quote Michael Vickers:

[RMAs] are major discontinuities in military affairs. They are brought about by changes in militarily relevant technologies, concepts of operations, methods of organization, and/or resources available. Relatively abruptly—most typically over two or three decades—they transform the conduct of war and make possible order-of-magnitude (or greater) gains in military effectiveness. They sharpen the advantage held by the strategic/operational offense and create enormous intertemporal differentials of capabilities between military regimes. A hierarchy of change links these revolutions with broader social, economic, and scientific transformation.[6]

In this sense, the transformation of the US military, initiated by the US Department of Defense (DoD) under Defense Secretary Donald Rumsfeld in the first half of the first decade of the 21st century, was originally intended to be a process of disruptive change and, by extension, was to herald a new phase of strategic innovation. The DoD's Office of Force Transformation (OFT)—established by Rumsfeld and initially headed by former Admiral Arthur Cebrowski, and intended to serve as the Pentagon's primary internal body for thinking about and implementing the RMA—defined US "defense transformation" as "a process that shapes the changing nature of military competition and cooperation through new combinations of concepts, capabilities, people, and organizations ...."[7] In short, defense transformation was nothing less than a fundamental shift in the manner in which US armed forces would conduct future warfare operations.

The elements of US defense transformation were several, but linked:

- A highly networked organism of command, control, communications, computing, intelligence, surveillance, and reconnaissance (C4ISR) systems, weapons, and platforms
- Improved, shared situational awareness, both of the immediate battlespace and even beyond
- More accurate, standoff engagement capacity
- Greater speed, agility, rapid deployability, and flexibility
- Jointness and interoperability[8]

Above all, the current US transformational/RMA model was inexorably linked to the emerging notions of *network-centric warfare* (NCW), sometimes also referred to as "network-enabled capabilities" or "network-based defense"—the operative word, of course, being "networked." According the NCW concepts, the ongoing revolution in information technologies (IT) has made possible significant innovation and improvement in the fields of sensors, seekers, data management, computing and communications, automation, range, and precision.[9] Correspondingly, NCW seeks to exploit these breakthroughs in information technology in order to achieve exponential improvements in battlefield knowledge, connectivity, and response.

Consequently, the US transformational model implied more a simple overlay of new technologies and new hardware over existing force structures; it entailed fundamental changes in military doctrine, operations, and organization. Hardware and technology are obviously crucial and primary components when it comes to transformation—they are fundamental building blocks in a modern, IT-based RMA centered on network-centric warfare and reconnaissance-strike complexes. Transformation was not supposed to be just a techno-fix, however, but rather was a fundamental (read, disruptive) change in the way the US military was supposed to fight—doctrinally, organizationally, and institutionally. Critically, too, therefore, it demanded fundamental changes in the ways that the US military would develop and procure critical military equipment, as well as the reform of the national and defense technological and industrial bases that were supposed to contribute to development and production of transformational systems. All this, in turn, required vision and commitment at the very top in order to develop the basic concepts of defense transformation to establish the crucial institutional and political momentum for implementing transformation, and to allocate the financial resources and human capital necessary for the task of implementation.[10] Obviously, therefore, US force transformation entailed much more than "mere" modernization.

The state of strategic innovation in the defense industry

If the RMA is all about *disruptive* strategic innovation, where does that leave the state-of-the-art in the global defense industry? In fact, recent experiences with radical innovation in the arms industry have not been very encouraging. Aerospace analyst Richard Aboulafia, for example, has made some interesting observations about the current state of aircraft development. In the first place, he noted that aircraft design has not changed much over the past several decades; especially in comparison with the promise of such futurist designs as the A-12 naval attack plane (once described as "the flying Dorito") or the commercial Piaggio Avanti, "planes today look basically like they did before those revolutionary jets arrived."[11] Moreover, he argued that

> Before it was unveiled, we thought the ATF [Advanced Tactical Fighter] would look completely new. The losing YF-23 contender was mildly exotic, but the F-22 (and F-35) are quite recognizable as conventional fighters. After the A-12 cancellation, [US Naval aviation] was saved by the F/A-18E/F, a derivative of a very conventional 1970s design.[12]

These innovation challenges have not just affected the fighter aircraft sector; take, for example, the unfortunate tale of the DDG-1000 *Zumwalt*-class destroyer. Few shipbuilding programs have been more fraught with setbacks than the USN's efforts to come up with a successor to the DDG-51 *Arleigh Burke*-class destroyer and the CG-47 *Ticonderoga*-class cruiser. In the 1990s, the US Navy (USN) undertook studies on a "new surface combatant for the 21st Century" (SC-21), which included both a new destroyer (DD-21) and cruiser (CG-21). Subsequent defense budget cuts later forced the navy to scale back this program to a destroyer-sized hull, designated the DD(X), which would be expanded to allow for a new cruiser, the CG(X). The DD(X) was later re-designated the DDG-1000 *Zumwalt*-class destroyer. The 16,000-ton DDG-1000 is a highly unique vessel, with a stealthy radar and sound signature, a "tumblehome" wave-piercing bow, electric-drive propulsion, an advanced gun system, several vertical launch system (VLS) tubes for Tomahawk, Evolved Sea Sparrow, and ASROC missiles (and possibly a railgun or free-electron laser, sometime in the future). Additionally, the ship was intended to be an epitome of network-centric warfare, with a highly automatic, integrated, and networked command and fire-control system.

The USN had hoped to acquire up to 32 ships in this class, but the DDG-1000 quickly ran into a number of problems. The US Government Accountability Office (GAO) determined that, out of 12 critical technologies, only four were considered "fully mature," while five technologies would "not demonstrate full maturity until after installation," and two others would "remain at a lower level of maturity" at least two years after the lead ship was launched.[13] At the same time, the cost of one *Zumwalt*-class destroyer had risen to at least $3.3 billion. Eventually, the number of *Zumwalt*-class destroyers was reduced from 24 ships to seven and finally to just three, while the CG(X) program (originally planned for 18 to 19 ships) was cancelled entirely. Instead, the USN will likely acquire additional (up to 24 more) DDG-51 class destroyers.

More often than not, therefore, tradition—at the expense of radical innovation—wins out over conceptual design. Edward Luttwak has argued that militaries (and by extension the global arms industry that supplies them) are "prisoners of tradition," and that consequently the "canonical weapons platforms and configurations of World War II have endured, despite all the new possibilities opened by technological advancement over the past six decades."[14] He added that

> [i]nstead of shaping new platforms and weapons configurations to fit today's information technology, communications, sensor and guidance equipment, we are shoving, cramming and molding such technology to fit the nooks and crannies of 1945-era platforms.[15]

For example, Luttwak noted that, according to modern combat aircraft design, every fighter jet has to have its own airborne radar—which not only makes such requirements very expensive (given the costs of miniaturization of electronics for such radar systems) but ignores the possibilities of other arrangements (such as "slaving" aircraft to non-integrated sensors and fire-control, which are ostensibly enabled by innovations in datalinks and other types of communications). Consequently, weapons systems such as aircraft, tanks, and surface combatants look a lot today like they did 50 or 75 years ago.

If basic conceptual designs for modern weapons systems haven't changed much, other potential technological innovations seem to have languished as well. Aboulafia has argued, for example, that stealth technology in aerospace has not yet really delivered on its promises. He notes that more than 20 years after the F-117 was unveiled, the US Air Force remains "the only air service in the world with stealth *and it's planning to operate non-stealthy F-15s and F-16s for decades to come.*"[16] The F-117 is already being retired, while the B-2 bomber was curtailed at just 21 aircraft and the F-22 at 189 fighters. The only very low observable (VLO) aircraft currently in production is the F-35 Joint Strike Fighter. The Europeans have no plans at the moment (and no money) to develop a fifth-generation fighter; and while the Russians tout the PAK FA and the Chinese the J-20 as VLO fighters, their stealthiness is still undetermined.

Even the central theme of the current IT-led RMA—network-centric warfare—seems to be faltering. The once-transformational promise of NCW now appears to have been downgraded to simply being better C4ISR. Take, for example, the US Army's Future Combat System, originally a $160 billion program to develop a fleet of 18 different versions of a single type of modularized light combat vehicle, both manned and unmanned, which in turn was to be linked together by a communications systems capable of providing real-time intelligence, command, and control to troops on the move. In 2009, the manned vehicles element of the FCS program, along with many of its integrated unmanned aerial vehicles (UAVs), were cancelled, and replaced by the Future Force Warrior, a much less ambitious program to expand ground forces' situational awareness through greatly improved individual soldier communications systems, sensors, and the like. Additionally, another key element of the US military's force transformation program, the Transformational Satellite Communications System (TSAT)—a secure, high-capacity satellite system that was supposed to serve as the space-based constituent of the Pentagon's Global Information Grid—was also eventually cancelled, while another program, the Joint Tactical Radio System (JTRS), was heavily scaled back.

Stagnation of strategic innovation in the global arms industry: Money and mindsets

The above examples are not intended to convey the impression that there is *no* strategic innovation still taking place. But at present, it seems that most militaries, and by extension, the global arms industry, are currently shifting away from *disruptive* innovation and towards more steady-state but low-level types of *continuous, sustaining* innovation. Why is this occurring?

In the first place, funding would appear to have a significant impact on individual nations' abilities to engage in disruptive strategic innovation when it comes to military programs. Quite simply, the types of disruptive technologies found in emerging weapons systems—particular those relating to the IT-led RMA—are increasingly too expensive for most countries to develop and integrate into their militaries. Most countries have relatively meager—generally just a few hundred millions or at most a few billions of dollars annually—that they can dedicate to military R&T. Consequently, the center of *defense-specific* innovation is increasingly shrinking, and to just one country, that is, the United States. This point is particularly apropos of Western Europe, where defense innovation—indeed, most technological innovation, it seems—is languishing.[17] Again, this is largely the result of funding: European military spending has been more or less stagnant since the end of the Cold War, that is, more than 20 years ago. Total Western European defense R&D spending is only around $12 billion, less than one-sixth that of the United States. Moreover, European military R&D is fragmented along national lines (France's defense R&D expenditures in 2009 were $5.4 billion, the largest in Europe, while Britain's was $3.9 billion; together, these two countries accounted for nearly four-fifths of all European military R&D spending), and more importantly, is generally allocated to duplicative, competing programs; consequently, the "buying power" of European military R&D spending is diluted and weakened, inhibiting technology development and defense innovation.

Europe's relative decline as a center of defense innovation is evident in the absence of new cutting-edge R&D programs. Aside from a handful of programs—such as the *Meteor* ramjet-powered, medium-range air-to-air missile, the *Visby*-class stealth corvette, or the Type-212 submarine (equipped with fuel cells for air-independent propulsion) there is little going on in the European defense industry that is not simply a replication of an existing US capability (e.g., the A400M transport plane). More critically, the European aerospace industry has no indigenous programs under even initial development for a next-generation fighter to compete with the US-designed F-35 Joint Strike Fighter. One possibility might be an unmanned combat aerial vehicle (UCAV), but so far, talk of a European UCAV has remained mostly that—talk.

Europe is not the only region to be so affected. Russia has been unable to move forward with its plans to develop a fifth-generation fighter aircraft due to a longstanding funding shortfall; consequently, it has been forced to partner with India, in hopes that New Delhi has the sufficient "deep pockets" to see this program to fruition. Japan's defense industry is at a similar standstill, given decades of declining procurement spending; armaments production has basically become a "job's program," and the country's earlier goals of defense industrial autarky (*kokusanka*) have largely been abandoned. Many other countries that aspire to becoming major arms producers—particularly India and South Korea—face similar problems of insufficient funding when it comes to underwriting disruptive innovation. *In general, therefore, many smaller arms-producing nations have remained imitators of Western technology, engaged, at most, in incremental and sustaining innovation.*

Even the United States is experiencing difficulties with maintaining its earlier pace of defense innovation, as potentially revolutionary programs such as the Future Combat System are cancelled or heavily scaled back, and as projects like the over-budget and heavily delayed F-35 suck up resources that could be dedicated to other developmental programs, such as a future bomber or a UCAV.[18] A recent study by the US Army reported that between 1995 and 2009 it spent more than $32 billion on programs that were subsequently cancelled. These projects included the Future Combat System ($19 billion spent over eight years), the Comanche reconnaissance-attack helicopter ($5.9 billion over ten years), the Crusader artillery system ($2.8 billion over eight years), and the Brilliant Antitank (BAT) "smart munition" for the ATACMS guided missile ($1.6 billion over 11 years).[19]

Beyond mere costs, it also appears that the global defense industry is not inclined to engage in radical innovation in the absence of proper guidance, support, and incentives from the user, that is, the military. In their study of the US defense industrial base and its efforts to supply the USN, Dombrowski, Gholz, and Ross have argued that, when it comes down to it, most military customers are simply not interested in disruptive innovation. Rather, they prefer sustaining innovations that, incidentally, already established suppliers (i.e., shipbuilders, defense electronics manufacturers, etc.) are capable of providing. Consequently, innovation will continue to be less traumatic than might be expected, both for users and producers.[20]

Mary Kaldor further expands this argument, asserting that military bureaucracies, being naturally "conservative" and operating according to "dominant scenarios," are not really comfortable with radically new technologies, since these "pose a risk for organizational survival."[21] Hence, the demand side of the procurement equation is already predisposed toward existing, proven technologies. Meanwhile, on the supply side, the prime contractors who undertake the development and manufacture of military equipment are, in response, basically engaged in a continuous process of product improvement, rather than disruptive innovation. But, she argues:

> These improvements, however, are severely constrained by the organizational rigidities of the armed forces. New technologies can only get through the innovation and integration stages if they conform to the requirements of the dominant scenario. So long as they are directed toward the improvement in the performance of missions that were established nearly 40 years ago and are currently defined through a set of performance characteristics or parameters (speed, payload, or protection) that have hardly changed, the technologies may be extremely radical in hardware terms—nuclear devices, directed energy, variable geometry aircraft, and so forth. *I have termed the contradictory conservative but dynamic form of technological change as baroque.*[22]

"Baroque" technological change, she adds, "represents 'improvements' to successive weapons systems which can pass through the phases of invention, innovation, and integration without disturbing the social organization of the users ...."[23]

Interestingly, some have argued that emphasizing sustaining over disruptive innovation may not be such a bad thing. Bracken, Brandt, and Johnson, in a paper written under the auspices of the US National Defense University, have argued that the US military is actually "overwhelmed" by new technologies and "the complexity of recent hardware," so much that it is actually adversely affecting the military's capabilities.[24] They start from the basis that defense procurement in the US military is still largely driven by a "legacy defense innovation system." This system was "built around the trade-off of increased performance for increased cost. As long as performance grew faster than cost, innovation paid off."[25] Unfortunately, they argue, this is rarely the case, and most military systems in the US arsenal (they generally exclude aerospace from this assertion) never achieve sufficient gains in capabilities to justify their rising costs. This problem is only exacerbated as technological complexity of new systems increases, as "the current system is so oriented to new production that it is choking the capacity to integrate these products into legacy forces."[26] The answer, according to the authors, is actually to move away from "the radical new product" solution and rather embrace upgrades and retrofits of existing weapons systems, as these are "easier to initiate, fund, and sustain."[27] In writing this, the authors are tacitly supporting the idea of moving from disruptive to continuing innovation as a strategic procurement policy—and, in fact, the Pentagon is already doing this, in the form of modernizing older systems such as the Multiple Launch Rocket System (MLRS) or converting dumb bombs into smart munitions via JDAM kits.[28]

Conclusions

So where does all this leave us? In the first place, it seems reasonable to assert that strategic innovation in the global arms industry, despite the transformational promise of the IT-led RMA, is not presently oriented toward disruptive, discontinuous, and revolutionary innovation. Rather, it would seem that sustaining innovation still predominates throughout much of the global defense industry. There is, of course, nothing wrong with such an approach. Continuous innovation can still produce amazing results eventually, and the "Brackensian model" of upgrades and retrofits may in fact be the smarter way to go.

More critically, however, this possible "lull" in disruptive strategic innovation on a more or less global scale may provide a pause or slowdown in the global process of defense technology development that would permit latecomer innovators and "fast followers" to draw nearer to the state of the art. This is particularly apropos in the case of China. China has been putting significant resources into its defense establishment, including increasing military expenditures at least *five-fold* (in real, i.e., after-inflation spending) over the past 15 years. Chinese military expenditures in 2011 totaled nearly $92 billion, making it the second largest defense spender in the world. More important, its military equipment budget has grown from around $3 billion in 1997 to around $30 billion today, and its defense R&D, although classified, probably approaches $6 billion annually. In other words, China simply has more money to throw at its defense development. At the same time, should the overall process of global defense innovation slow, then China might have an opportunity to catch up. Certainly in its pursuit of a fifth-generation fighter aircraft (e.g., the J-20 and the J-31 protoypes), it is poised to overtake Europe in this one particular area. Overall, while China may not supplant Europe as a defense innovator, it could at least be gaining capacities to match Europe in certain niche areas, although this may take several years, if not decades, to be realized.

In sum, it is not that strategic innovation is going away, of course, but it certainly will not be like what many had promised back in the 1990s that it would be. For better or for worse, it would appear that more conservative types of sustaining continuous innovation are edging out more radical, disruptive innovation processes in the global arms industry. As argued earlier, there is nothing wrong with such an approach, and, indeed the process of sustaining innovation may be the smarter path. At the same time, emphasizing sustaining over disruptive innovation—particularly when viewed against the backdrop of all the heady excitement aroused by the promise of the IT-led RMA in the 1990s and 2000s—seems prosaic and uninspiring in comparison. One thing is for certain: the future ain't what it used to be.

Notes

1. Peter J. Dombrowski, Eugene Gholz, and Andrew L. Ross, "Selling Military Transformation: The Defense Industry and Innovation," *Orbis*, Summer 2002, p. 527. See also Clayton M. Christensen, *The Innovator's Dilemma* (New York: Harper Business, 2000).
2. Dombrowski, Gholz, and Ross, "Selling Military Transformation," p. 527.
3. "Transformed: A Survey of the Defense Industry," *The Economist*, 20 July 2002, p. 7.
4. Andrew Krepinevich, "From Cavalry to Computer: The Pattern of Military Revolutions," *The National Interest*, Fall 1994, p. 30.
5. Richard O. Hundley, *Past Revolutions, Future Transformations* (Santa Monica, CA: RAND, 1999), quoted in "Transformed: A Survey of the Defense Industry," p. 7.
6. Michael J. Vickers, "The Revolution in Military Affairs and Military Capabilities," in Robert L. Pfaltzgraff Jr and Richard H. Schulz Jr, eds., *War in the Information Age: New Challenges for U.S. Security Policy* (Washington DC: Brassey's, 1997), p. 30.
7. U.S. Department of Defense, Office of Force Transformation (OFT), *Elements of Defense Transformation* (Washington, DC: Office of the Secretary of Defense, October 2004), p. 2.
8. Andrew Ross, *Transformation: What Is It? What Does It Mean for Industry?*, PowerPoint briefing presented at the conference on "Defense Transformation in the Asia-Pacific: Meeting the Challenge," Honolulu, Hawaii, March 30–April 1, 2004, p. 2.
9. Center for Strategic and Budgetary Assessments (CSBA), *The Emerging RMA* (Washington, DC: CSBA, undated), downloaded from the CSBA website: www.csbaonline.org/2Strategic_Studies/2Emerging_RMA/Emerging_RMA. html, accessed 2 April 2014.
10. Center for Strategic and Budgetary Assessments, *Transformation Strategy* (Washington, DC: CSBA, undated), downloaded from the CSBA website: www.csbaonline.org/2Strategic_Studies/3Transformation_Strategy/Transform ation_Strategy.html.
11. Richard Aboulafia, *October 2010 Newsletter*, http://www.richardaboulafia. com/shownote.asp?id=323, accessed 2 April 2014.
12. Aboulafia, *October 2010 Newsletter*.

13. U.S. Congress, U.S. Government Accountability Office, *Defense Acquisitions: Assessments of Selected Weapon Programs* (Washington, DC: GAO, March 2009), pp. 69–70.
14. Edward N. Luttwak, "Breaking the Bank: Why Weapons Cost So Much," *The American Interest*, Autumn (September/October) 2007, p. 51.
15. Luttwak, "Breaking the Bank," p. 51.
16. Aboulafia, *October 2010 Newsletter*. Italics added.
17. Brian Palmer, "Good at Wine, Bad at Computers: Why Does Europe Suck at Technological Innovation?" *Slate*, 8 June 2011, http://www.slate.com/id/2296547, accessed 2 April 2014.
18. Aboulafia, *October 2010 Newsletter*.
19. Marjorie Censer, "Army Report: Military Has Spent $32 Billion Since '95 on Abandoned Weapons Programs," *Washington Post*, May 27, 2011.
20. Dombrowski, Gholz, and Ross, "Selling Military Transformation," pp. 523–536.
21. Mary Kaldor, "The Weapons Succession Process," *World Politics*, Vol. 38, No. 4 (July 1986), pp. 406–418.
22. Kaldor, "The Weapons Succession Process," p. 412. Italics added.
23. Kaldor, "The Weapons Succession Process," p. 415.
24. Paul Bracken, Linda Brandt, and Stuart E. Johnson, "The Changing Landscape of Defense Innovation," *Defense Horizons No. 47*, Center for Technology and National Security Policy, National Defense University, July 2005.
25. Bracken, Brandt, and Johnson, "The Changing Landscape of Defense Innovation," p. 3.
26. Bracken, Brandt, and Johnson, "The Changing Landscape of Defense Innovation," p. 7.
27. Bracken, Brandt, and Johnson, "The Changing Landscape of Defense Innovation," p. 2.
28. Bracken, Brandt, and Johnson, "The Changing Landscape of Defense Innovation," pp. 2, 7.

Index